# 14th Pacific Rim Bio-Based Composites Symposium 2018

## Bio-Based Composites for Better Green Life Cycles

IOP Conference Series: Materials Science and Engineering
Volume 593



Makassar, Indonesia
29 - 31 October 2018

**Editor:**

**Suhasman**

ISBN: 978-1-5108-9465-5
ISSN: 1757-8981

**Printed from e-media with permission by:**

Curran Associates, Inc.
57 Morehouse Lane
Red Hook, NY 12571

CURRAN ASSOCIATES INC. proceedings.com

**Some format issues inherent in the e-media version may also appear in this print version.**



Institute of Physics
Dirac House, Temple Back
Bristol BS1 6BE UK

Phone: 44 1 17 929 7481
Fax: 44 1 17 920 0979

techtracking@iop.org

**Additional copies of this publication are available from:**

Curran Associates, Inc.
57 Morehouse Lane
Red Hook, NY 12571 USA
Phone: 845-758-0400
Fax: 845-758-2633
Email: curran@proceedings.com
Web: www.proceedings.com

# Table of contents

## Volume 593

**The 14th Pacific Rim Bio-Based Composites Symposium**

**29–31 October 2018, South Sulawesi Province, Indonesia**

**Accepted papers received: 22 July 2019**
**Published online: 19 August 2019**

### Preface



### Papers

IOP Conf. Series: Materials Science and Engineering **593** (2019) 011001 doi:10.1088/1757-899X/593/1/011001

# Preface

After Rotorua, Vancouver, Kyoto, Bogor, Canberra, Portland Oregon, Nanjing, Kuala Lumpur, Rotorua, Alberta, Shizuoka, Beijing, and Concepcion, became the previous hosts of the Pacific Rim Bio-based Composite Symposium, Hasanuddin University, Makassar-Indonesia has been given an honour become the host for the symposium in the year of 2018.

It is an honor for us to present some of the papers that have been reviewed and presented at the symposium in this proceeding, both through oral presentation and poster presentation. The theme of this symposium is "**Bio-based composites for better green life cycles**" where the green life cycle is implemented through utilization green material and its processing. Scientists in the field of bio-based composites have been continuing highlight recent research in wood-based panel and composites. This symposium aims to be forum to exchange opinions, form research partnership and promote new research and development in bio-based composite. In addition to specific topics related to biocomposite and wood products, this symposium also provides media to present papers relating to forest products and forestry. The symposium covers topics of aspect biocomposite materials, fundamental and advance biocomposite processing, adhesive and bonding, polymer composite and bioplastic product, modified wood composite, perspective and challenge of composite product industrially and environmentally, forestry, forest product, and woody plant research which are presented in this proceeding.

On behalf of the editors, we would like to thank you for the presenter in the symposium and also all the attendants. The participants are from Japan, Republic of Korea, Taiwan, Rumania, Philippines, India, and Indonesia. In particular, we would like to express our gratitude for the cooperation of the authors in preparing the paper contained in this proceeding.

We would like to extend our gratitude to the Ministry of Research Technology and Higher Education through Directorate General of Higher Education, President of Hasanuddin University for the support and sponsorship, our distinguish keynote speakers, Prof. Shigehiko Suzuki from Shizuoka University, Prof. Byung Dae Park from Kyungpook National University and Prof. Bambang Subiyanto from Indonesian Institute of Sciences.

The symposium could not be held without the support and cooperation of other sources. Finally, our gratitude goes to the Faculty of Forestry Dean Hasanuddin University, Prof. Yusuf Sudo Hadi, and all the international committee board, local committee member, and students as volunteers in this symposium.

Makassar, October 2018

Suhasman

Chief Editor

## Acknowledgement

The committee symposium would like to express our gratitude to the Ministry of Research-Technology and Higher Education of Indonesian Republic through Directoral General of Higher Education, and Hasanuddin University for the support and sponsored this Pacific Rim Bio-based Composite Symposium and the proceeding.

IOP Conf. Series: Materials Science and Engineering **593** (2019) 011003 doi:10.1088/1757-899X/593/1/011003

**List of Partcipant The 14th Pacific Rim Bio-Based Composites Symposium**
**29-31, October 2018, Swiss-BelHotel, Makassar**

| No | Name | Institution | Country | Email |
|---|---|---|---|---|
| 1 | Bambang Subiyanto | Indonesian Institute of Sciences, Indonesia | Indonesia | dr.bambang.subiyanto@gmail.com |
| 2 | Shigehiko Suzuki | Faculty of Agriculture, Shizuoka University, Japan | Japan | suzuki.shigehiko@shizouka.ac.jp |
| 3 | Byung-Dae Park | Kyungpook National University, Republic of Korea | Korea | byungdae@knu.ac.kr |
| 4 | Kenji Umemura | Kyoto University, Japan | Japan | umemura@rish.kyoto-u.ac.jp |
| 5 | Yusuf Sudo Hadi | Bogor agricultural University | Indonesia | yshadi@indo.net.id |
| 6 | Andi Sri Rahayu Diza | Bogor agricultural University | Indonesia | ayudiza@gmail.com |
| 7 | Andi Tri Lestari | Forestry Department, Mataram University, Indonesia | Indonesia | atlestari@unram.ac.id |
| 8 | Arif Nuryawan | North Sumatera of University, Indonesia | Indonesia | arifnury@yahoo.com |
| 9 | Deazy Trisatya | Forest Product Research and Development Centre, Ministry of Environment and Forestry Republic of Indonesia | Indonesia | drtrisatya@gmail.com |
| 10 | Emilia-Adela Salca | Transilvania University of Brasov | Rumania | emilia.salca@unitbv.ro |
| 11 | Feng-Cheng Chang | School of Forestry and Resource Conservation, National Taiwan University | Taiwan | fcchang@ntu.edu.tw |
| 12 | Firda Aula Syamani | Indonesian Institute of Sciences, Indonesia | Indonesia | firda.syamani@biomaterial.lipi.go.id |

| | | | | |
|---|---|---|---|---|
| 13 | HSIN-CHEN | School of Forestry and Resource Conservation, National Taiwan University | Taiwan | christy850114@gmail.com |
| 14 | Ignasia Maria Sulastiningsih | Forest Product Research and Development Centre, Ministry of Environment and Forestry Republic of Indonesia | Indonesia | ignasiasulastin@gmail.com |
| 15 | Iwan Risnasari | North Sumatera of University, Indonesia | Indonesia | iwanrisnasari@gmail.com |
| 16 | Jamaluddin Malik | Forest Product Research and Development Centre, Ministry of Environment and Forestry Republic of Indonesia, | Indonesia | jmalik.unimelb2012@gmail.com |
| 17 | Kazhusige Murayama | United Graduate School of Agricultural Science, Gifu University, Shizuoka, Japan | Japan | v6103009@edu.gifu-u.ac.jp |
| 18 | Kazuaki Ota | Faculty of Agriculture, Shizuoka University, Japan | Japan | ota.kazuaki.17@shizuoka.ac.jp |
| 19 | Manojit DAS | PT. Arjuna Utama Kimia | India | mdas@aruki.co.id |
| 20 | Muryani Arsal | Muhammadiyah University of Makassar, Indonesia | Indonesia | muryaniarsal@gmail.com |
| 21 | Nanang Masruchin | Indonesian Institute of Sciences, Indonesia | Indonesia | masruchin@biomaterial.lipi.go.id |
| 22 | Nindya Fertikasari | Faculty of Agriculture, Shizuoka University, Japan | Japan | nindyaferrtikasari@gmail.com |
| 23 | Ragil Widyorini | Gadjah Mada University | Indonesia | rwidyorini@ugm.ac.id |
| 24 | Rahma Nur Komariah | Kyoto University, Japan | Japan | rahma_komariah@rish.kyoto-u.ac.jp |
| 25 | Rahmawaty | North Sumatera of University, Indonesia | Indonesia | rahmawaty@usu.ac.id |

| | | | | |
|---|---|---|---|---|
| 26 | Ronniel Manalo | University of the Philippines Los Baños Philippines | Filipina | rdmanalo@up.edu.ph |
| 27 | Seok un Jo | Chungbuk National University, Korea | Korea | jo18041@naver.com |
| 28 | Sujito | Faculty of Mathematics and Sciences, University of Jember | Indonesia | sujito.fmipa@unej.ac.id |
| 29 | Sung Phil Mun | Chungbuk National University, Korea | Korea | msp@jbnu.ac.kr |
| 30 | Tamaryska Setyayunita | Gadjah Mada, University | Indonesia | ryskasetya@gmail.com |
| 31 | Tito Sucipto | North Sumatera of University, Indonesia | Indonesia | titomedan@yahoo.com |
| 32 | Yoichi Kojima | Faculty of Agriculture, Shizuoka University, Japan | Japan | kojima.yoichi@shizuoka.ac.jp |
| 33 | Zakiah Uslinawaty | Haluoleo, University | Indonesia | zakiahuslinawati@gmail.com |
| 34 | Muhammad Navis Rofii | Gadjah Mada, University | Indonesia | navis_r@ugm.ac.id |
| 35 | Jung Myoung Lee | Kyungpook National University, Republic of Korea | Korea | jmylee@knu.ac.kr |
| 36 | Lukmanul Hakim Zaini | Bogor agricultural University | Indonesia | lukmanhz@apps.ipb.ac.id |
| 37 | Dahlan Tahir | Faculty of Mathematics and Science, Hasanuddin University | Indonesia | d.tahir@fmipa@.unhas.ac.id |
| 38 | Sanghee Sur | Korean Society Wood of Science Technology | Korea | xiang-xi@hanmail.net |
| 39 | Yun So Young | Gyeongju CVB | Korea | soyoung@hicogw.kr |
| 40 | Gyu Seong Han | Chungbuk National University, Korea | Korea | |
| 41 | Young Ho Seo | Gyeongju CVB | Korea | yhseo@crowncity.kr |

IOP Conf. Series: Materials Science and Engineering **593** (2019) 011003 doi:10.1088/1757-899X/593/1/011003

| | | | | |
|---|---|---|---|---|
| 42 | Sang Woo Hoh | Korean Forest Service | Korea | nohsw330@korea.kr |
| 43 | Budiaman | Hasanuddin University | Indonesia | lintassapi@yahoo.com |
| 44 | Beta Putranto | Hasanuddin University | Indonesia | betaputranto@yahoo.co.id |
| 45 | A.Sadapotto | Hasanuddin University | Indonesia | sadapotto70@gmail.com |
| 46 | Makarennu | Hasanuddin University | Indonesia | nunumakkarennu@gmail.com |
| 47 | Rohny. S. Maail | Patimura University | Indonesia | rohny_maail.yahoo.com |
| 48 | Gusmiaty | Hasanuddin University | Indonesia | umyhody@ymail.com |
| 49 | Supriyanto | Nature Conservation Agency, Makassar | Indonesia | |
| 50 | Khairan arsyad | Nature Conservation Agency, Makassar | Indonesia | motojunkiez@gmail.com |
| 51 | Nisikido Kunihari | Osaka | Japan | nishikido@fa.sekisuihouse.co.jp |
| 52 | Dede Hermawan | Bogor agricultural University | Indonesia | dedemjmr@yahoo.co.id |
| 53 | Abdurachman | Forest Products Research and Development Centre, Ministry of Environment and Forestry | Indonesia | man_p3hh@yahoo.com |
| 54 | Wa Ode Muliastuty Arsyad | Forest Product Research and Development Center, Ministry of Environment and Forestry | Indonesia | waodemuliastuty@gmail.com |
| 55 | Apri Heri Iswanto | North Sumatera of University, Indonesia | Indonesia | apriheri@yahoo.com |
| 56 | Siti Fatimah | Hasanuddin University | Indonesia | fatimahsamsul02@gmail.com |
| 57 | Nurhayati Hadjar | Haluoleo, University | Indonesia | nurexdelaforest@gmail.com |

IOP Conf. Series: Materials Science and Engineering **593** (2019) 011003 doi:10.1088/1757-899X/593/1/011003

| | | | | |
|---|---|---|---|---|
| 58 | Rosmarlinasiah | Haluoleo, University | Indonesia | rosmarlinasiah.cuma@gmail.com |
| 59 | Bakri | Hasanuddin University | Indonesia | bakrifhut@unhas.ac.id |
| 60 | Siti Halimah | Hasanuddin University | Indonesia | sitih5h.82@gmail.com |
| 61 | Kazuaki Takahashi | Osaka | Japan | kazuakit@gz.sekisuihouse.co.jp |
| 62 | Suhasman | Hasanuddin University | Indonesia | suhasman@yahoo.com |
| 63 | Sahriyanti Saad | Hasanuddin University | Indonesia | arie_tanti@yahoo.com |
| 64 | Andi Detti Yunianti | Hasanuddin University | Indonesia | dettiyunianti70@yahoo.com |
| 65 | Syahidah | Hasanuddin University | Indonesia | syahidah@unhas.ac.id |
| 66 | Ira Taskirawati | Hasanuddin University | Indonesia | tasqira@gmail.com |
| 67 | Astuti Arif | Hasanuddin University | Indonesia | astuti_arif@unhas.ac.id |
| 68 | Agus salim | Hasanuddin University | Indonesia | agussalim.madjid@yahoo.co.id |
| 69 | Musrizal Muin | Hasanuddin University | Indonesia | musrizal@yahoo.com |
| 70 | Desi Patandianan | Hasanuddin University | Indonesia | desipatandianan11@gmail.com |
| 71 | S. A. Rizky Awalia | Hasanuddin University | Indonesia | sa.reskiawalia@gmail.com |

IOP Conf. Series: Materials Science and Engineering **593** (2019) 011003 doi:10.1088/1757-899X/593/1/011003

# COMMITTEE

**International Steering Committee:**

- James Hague – Australia
- John Wolodko – Canada
- Changtong Mei – China
- Fu Feng – China
- Philip Humphrey – United States
- Shigehiko Suzuki – Japan
- Warren Grigsby – New Zealand
- Yusuf Sudo Hadi – Indonesia
- Paridah MD. TAHIR – Malaysia
- Chunping Dai – Canada
- Roger Rowell – United States (Chair)
- Alex Berg – Chile

**Organizing Committee:**

- Chair: Suhasman

  Members:
- Sahriyanti Saad
- Andi Detti Yunianti
- Ira Taskirawati
- Musrizal Muin
- Astuti Arif
- Syahidah
- Agussalim

1. Rector of Hasanuddin University, Prof Dwia Aries Tina Pulubuhu, open the symposium

2. Photo season

IOP Conf. Series: Materials Science and Engineering **593** (2019) 011004 doi:10.1088/1757-899X/593/1/011004

3. Keynote Speaker – Prof. Dr. Shigehiko Suzuki. Shizuoka University Japan

4. Oral Presentation

5. Poster Presentation

6. Prof Byung Dae Park – Presenting the next Biocomp 2020 in Republic of Korea

IOP Conf. Series: Materials Science and Engineering **593** (2019) 011004 doi:10.1088/1757-899X/593/1/011004

7. Giving the stick to the next committee organizer

8. Gala dinner for the participants

IOP Conf. Series: Materials Science and Engineering **593** (2019) 011005 doi:10.1088/1757-899X/593/1/011005

# Peer review statement

All papers published in this volume of *IOP Conference Series: Materials Science and Engineering* have been peer reviewed through processes administered by the proceedings Editors. Reviews were conducted by expert referees to the professional and scientific standards expected of a proceedings journal published by IOP Publishing.

# Characteristics and Morphology of Biodegradable Plastics from Oil Palm Waste for Agricultural and Forestry Applications

**I Risnasari[1,2*], A Nuryawan[1], Delvian[1], Ridwansyah[3], K Sitompul[1], Affandi[1] and W Pulungan[1]**

[1]Faculty of Forestry, University of Sumatera Utara
Jl. Tridharma Ujung No. 1 Medan, Sumatera Utara 20155, Indonesia
[2]Center of Excellence for Natural Resources-Based Technology, University of Sumatera Utara, Medan-Sumatera Utara 20155, Indonesia
[3]Faculty of Agriculture, University of Sumatera Utara

*Email: iwanrisnasari@gmail.com

**Abtsract.** The use of conventional plastics in human life is very diverse, one of which is as a seedling container (polybags) for agriculture and forestry crop. Although the use of polybags is felt to make it easier for humans in the nursery but causes severe hazards to the environment. This research will synthesize biodegradable plastic from oil palm waste, in the form of cellulose and starch from the stem. The parameters that will be measured from this study are mechanical and thermal properties, as well as their morphology.

## 1. Introduction

Plastic composite products are widely circulating in the community in various uses, such as household furniture, kitchen appliances, automotive components to food packaging. These products are produced using base oil from non-renewable materials. In its use, plastic goods will produce plastic waste that cannot be decomposed by decomposing microorganisms in nature (non-biodegradable). Plastic products have caused environmental problems, namely accumulation in large quantities in nature [1]. One of the contributors of plastic waste in nature is in planting activities, especially in the needs of the agricultural and forestry sector planting media in the form of polybags. When planting seeds in the field, usually the plastic is left just like that so that there is accumulation in nature. Often polybags used can cause damage to the root system, thus affecting subsequent growth in the field (Noraskhin and Ibrahim, 2009; [2].

## 2. Materials and Methods

The materials used include palm oil plantations of PTPN II Tanjung Morawa plantation, glycerol, vinegar acid, and aquades. Making starch from palm oil is done by splitting the oil palm stem and then separating the hard skin and the spout [3]. The sponge is shaved to become wood powder. Wood powder plus water then squeezed then filtered with filter cloth. The pulp is set aside as cellulose fiber for plastic fillers later, while starchy water precipitated for 3 hours, then wet starch was produced.The wet starch was washed by adding water and deposited for 3 hours then the wet starch was dried in an oven at a temperature of 500 C until the starch moisture content was ± 10%.

IOP Conf. Series: Materials Science and Engineering **593** (2019) 012001 doi:10.1088/1757-899X/593/1/012001

Making biodegradable plastic is done by dissolving palm starch in distilled water with a ratio of 1:20. Then glycerol and vinegar acid were added, then rotated with magnetic stirrer at a temperature of 800C to form a gel. The plastic formation is done by the film casting method and in the oven at 600C. As a comparison, a plastic film from tapioca starch was made.

The crystallinity of composite films was measured using X-ray diffraction / XRD (Shimadzu XRD-7000 MaximaX, Japan) at 40 kV and 30 mA, angles 10 ° -40 ° with Ni CuKα radiation filters (λ = 1. 54060 Å). Characterization of mechanical properties (tensile strength, tensile modulus, and broken elongation) of composite films was carried out based on ASTM D882-75b (Standard Test Method for Tensile Properties of Thin Plastic Sheeting) at crosshead speeds of 50mm / min.

## 3. Result and Discussion

Figure 1 shows that it has been successfully synthesized biodegradable plastic films from palm starch as well as plastic films from commercial starch (tapioca) which have elastic properties. This is because oil palm has a very high starch content (above 95%), which is 96%, almost equals tapioca starch content of 96.99%, so it is very good as a raw material for modified starch products [3]. Tapioca starch-based plastic has a clearer and more transparent visual appearance, while plastic from palm starch is more opaque and slightly yellowish in color. This is because palm starch has a lower degree of whiteness and paste clarity than tapioca starch, thus affecting the color and clarity of the plastic film produced.

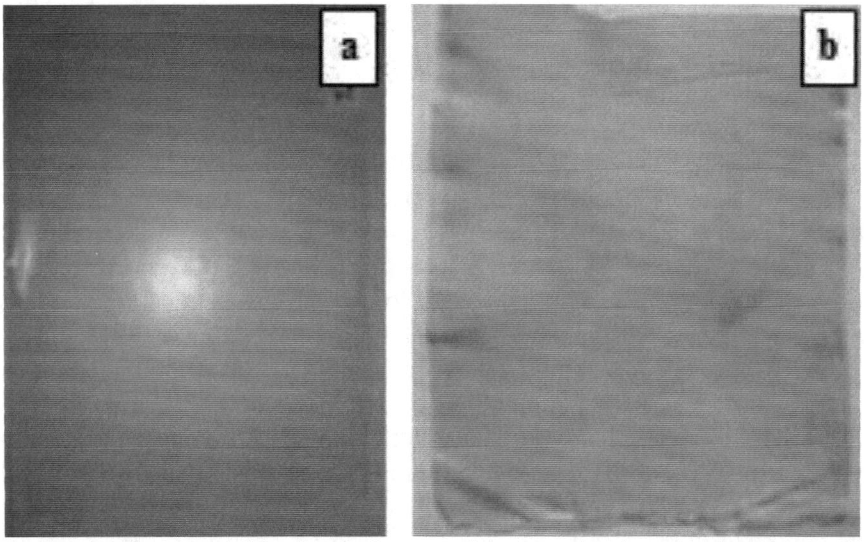


**Figure 1.** Biodegradable plastic from tapioca starch (a) and palm starch (b).

The surface morphology of biodegradable plastic film in Figure 2 shows that the plastic from palm starch has a surface that is not porous and looks finer than the plastic surface of tapioca starch. Oil palm starch has a higher final viscosity than commercial starch so that the palm starch becomes younger experiencing retrogradation [3]. During the retrogradation process amylose crystallizes so that the structure is compact and resistant to hydrolysis [4].

IOP Conf. Series: Materials Science and Engineering **593** (2019) 012001 doi:10.1088/1757-899X/593/1/012001

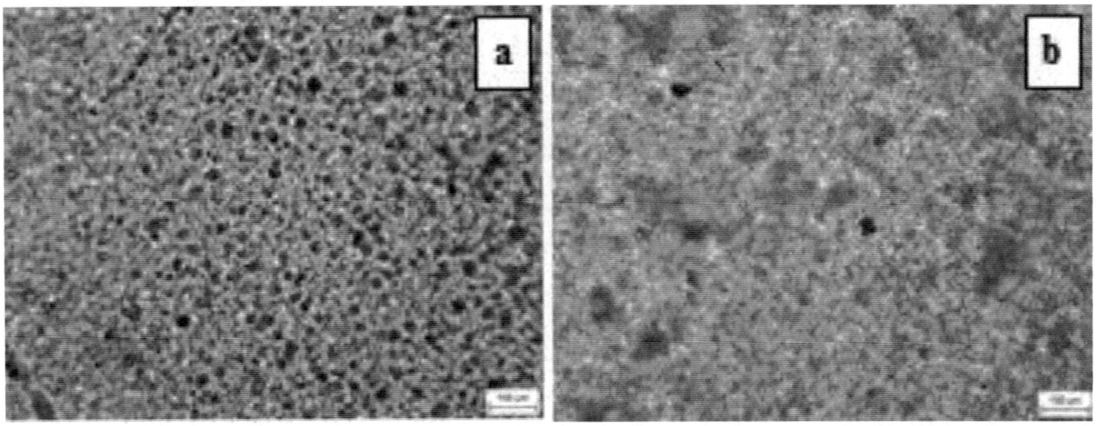


**Figure 2.** Plastic morphology Biodegradable plastic from tapioca starch (a) and palm starch (b), 10x magnification.

Figure 3 shows the value of water uptake by plastic based on palm starch is lower than tapioca-based plastic. Amylose content in palm starch was lower than tapioca starch, while amylopectin levels in palm starch were higher than tapioca starch [3]. The main components of edible film compiler are three groups: hydrocolloid, fat, and composites [5]. One of the main ingredients used in the manufacture of edible films is starch which belongs to the hydrocolloid group, which is an easy-to-obtain material, inexpensive, and of various types in Indonesia. Low levels of amylose and high amylopectin can facilitate the starch gelatinization process because it can reduce the solubility of starch in water so that starch can only expand in hot water which is needed in the starch gelatinization process. With high amylopectin levels, much space is available so that this space will be filled by mixing biopolymers.

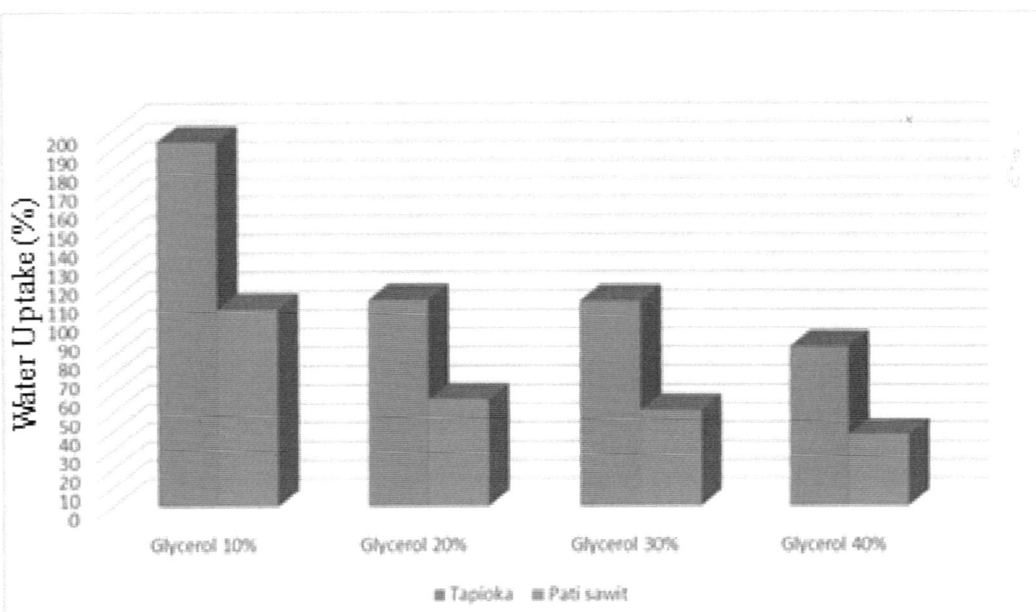


**Figure 3.** Water Absorption Percentage Graph by Plastic Made from Tapioca and Palm Starch

The results of testing with XRD showed that the degree of crystallinity in palm starch was higher than the crystallinity of tapioca starch, which was 31.51% and 27.78%, respectively. Higher crystallinity indicates a lower amorphous region so that the structure is more crystalline and stable. This will have an impact on the nature of the plastic that will be produced.

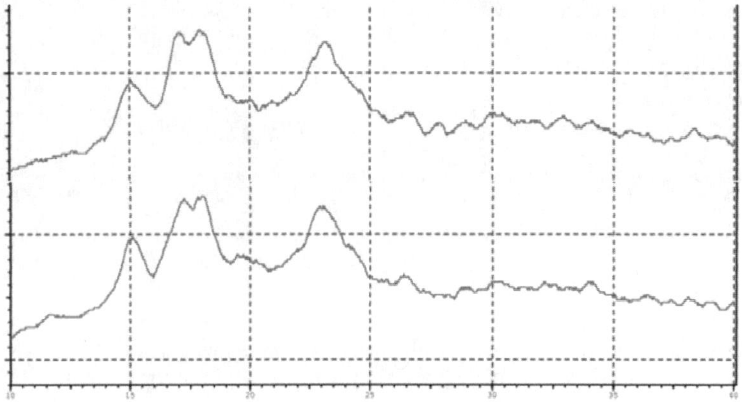

**Figure 4.** XRD Graph of Tapioca and Palm Starch-Based Plastics

The mechanical properties of starch-based degradable plastics are shown in Figure 5. Overall, it was shown that plastics made from palm starch had better mechanical properties than those made from tapioca starch. This is caused by the higher crystallinity of palm starch and the effect of amylose and amylopectin levels. Palm starch had lower amylose content than tapioca starch [4]. While the amylopectin content of palm starch is higher than the level of amylopectin in tapioca starch. Low levels of amylose and high amylopectin could facilitate the starch gelatinization process because it can reduce the solubility of starch in water so that starch can only expand in hot water which is needed in the starch gelatinization process [6]. With high amylopectin levels, much empty space is available so that this empty space will be filled by mixing biopolymers.

IOP Conf. Series: Materials Science and Engineering **593** (2019) 012001 doi:10.1088/1757-899X/593/1/012001

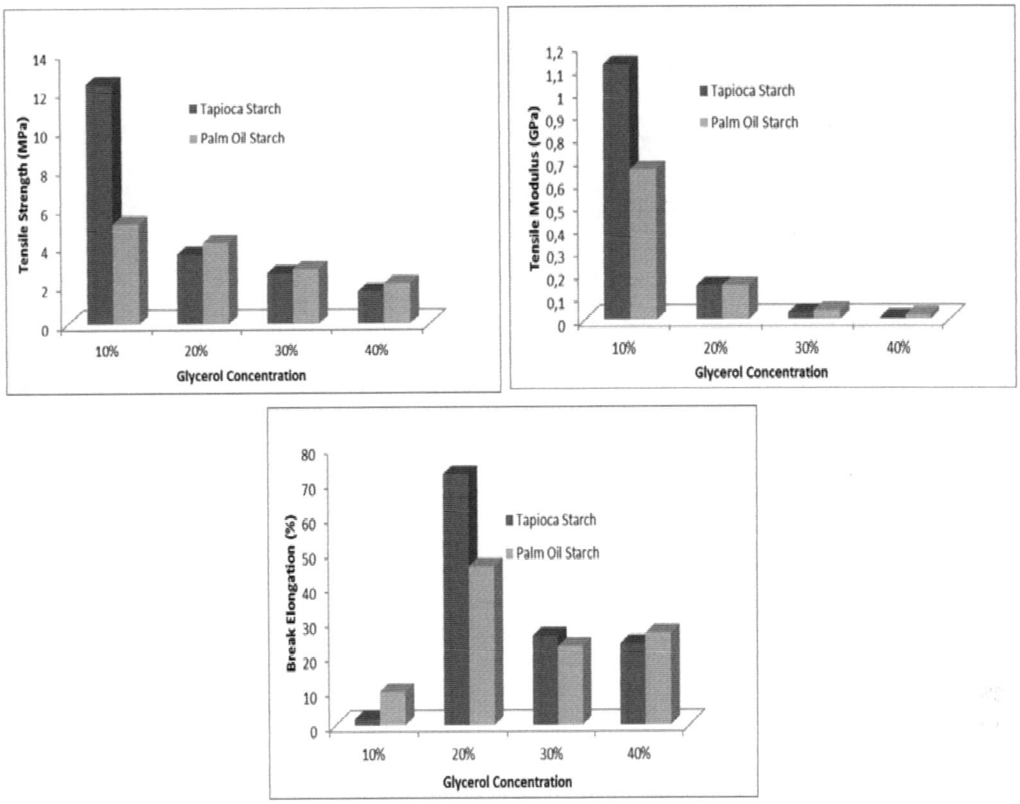


**Figure 5.** Graph of Tensile Strength, Tensile Modulus and Break Elongation in Tapioca and Palm Starch-Based Plastics.

## 4. Conclusion

This study shows that there are various types of glycerol from bioplastic sheets which will be affected by water uptake and tensile strength properties. The results obtained showed that the tapioca starch-based bioplastic sheets had the highest percentage in water uptake and the palm oil starch-based bioplastics sheets had the lowest water absorption. In the mechanical properties of some biodegradable plastics from oil palm starch has better properties, this is due to the higher crystallinity of palm starch and the influence of amylose and amylopectin levels.

## References

[1] Liew K C and Khor L K 2013 Effect of Different Ratios of Bioplastic to Newspaper Pulp Fiber on the Weight Loss of Bioplastic Pot. *Journal of King Saud University-Engineering Sciences* **27** pp 137–141

[2] Muriuki J K, Kuria A W, Muthuri C W, Mukuralinda A, Simons A J and Jamnadass R H. 2013 Testing Biodegradable Seedling Containers as an Alternative for Polythene Tubes in Tropical Small-Scale Tree Nurseries. Small-scale Forestry, Springer

[3] Ridwansyah, Nasution M Z, Sunarti T C and Fauzi A M 2007 Characteristics of Physico-Chemical Properties of Palm Oil Starch – in Bahasa *Jurnal Teknologi Industri Pertanian* **17**(1) pp 1–6

[4] Interpares P, Haryadi and Cahyanto M N 2015 Effect of Retrogradation on the Making of Corn Starch Vermicelli Against the Physicochemical Characteristics of Prebiotic Products and Activities – in Bahasa *Agritech* **35** (2) pp 192–199

[5] Rodriguez M, Oses J, Ziani K and Mate J I 2006 Combined Effect Of Plasticizer And Surfactants On The Physical Properties Of Starch Based Edible Films *Journal of Food Research International* **39** pp 840–846

[6] Setiani S, Tety S dan Rahmidar L 2013 Preparation and Characterization of Edible Films from Starch-Chitosan Starch Poly. *Valensi* **3** (2) pp 100–109

**Acknowledgments**

The authors would like to thank the Research Institutions, University of Sumatera Utara for its financial support by the TALENTA Research Implementation Contract of the University of Sumatera Utara, 2018 (Grant No: 2590 / UN.1.R / PPM / 2018 March 16, 2018).

IOP Conf. Series: Materials Science and Engineering **593** (2019) 012002 doi:10.1088/1757-899X/593/1/012002

# Some important properties of strandboard manufactured from andong bamboo (*Gigantochloa pseudoarundinacea*)

**I M Sulastiningsih[1*], D A Indrawan[1] and D R Trisatya[1]**

Forest Products Research and Development Centre
Research, Development and Innovation Agency
Ministry of Environment and Forestry, Republic of Indonesia
Jl. Gunung Batu No. 5, Bogor, Indonesia. Telp. (0251) 8633378

*Email :ignasiasulastin@gmail.com

**Abstract**. Andong bamboo (*Gigantochloa pseudoarundinacea*) is one of the most important bamboo species widely planted in West Java, Indonesia, and play a prominent role in the rural economy. The bottom and the middle parts of Andong bamboo culms are suitable for laminated bamboo lumber raw material. To improve the efficiency of bamboo utilization, the upper part of Andong bamboo culms were use as raw material for strandboard production. The objective of the study was to determine the effect of strand length and specific pressure on the properties of bamboo strandboard. Laboratory scale of bamboo strandboards were fabricated using three different strand lengths (75, 100, and 150 mm), two different specific pressures applied (25 kg/cm$^2$ and 30 kg/cm$^2$) and glued with phenol formaldehyde with the targeted density of 0.75 g/cm$^3$. Some physical and mechanical properties of bamboo strandboards were determined. Results showed that the average physical properties of bamboo srandboards were 9.3% (moisture content), 0.76 g/cm$^3$ (density), 3.1% (thickness swelling), 1.6% (wide expansion), and 25.1% (water absorpsion). The average mechanical properties of bamboo srandboards were 760.5 kgf/cm$^2$ (MOR), 92.3 10$^3$kgf/cm$^2$ (MOE), 5.6 kgf/cm$^2$ (IB), and 56.5 kgf (screw holding power). Some physical and mechanical properties of bamboo strandboard were significantly affected by strand length and specific pressure.

## 1. Introduction

Indonesia as one of the highest diversity in the world, consist of 160 bamboo species which consists of introduced species (38 species), origin species (122 species). Only 76 species (about 50% of the total species) are used by local people for their daily life, but only 9-10 species are commonly or commercially used in Java and Bali. From the west, bamboo has bigger diameter than in the east, however, bamboo in the highland has smaller diameter than bamboo in the lowland. In the western of Indonesia the genus *Dendrocalamus*, *Gigantochloa* and *Bambusa* grow abundantly compare to the eastern part [1]. Bamboo in Indonesia is one of the most important non timber forest products and has long been recognized as a multi-purpose plant. Bamboo has the potential to be an alternative to furniture and housing materials due to its ability to grow fast in various soils with desirable properties.

In two decades research on bamboo utilization as raw material for laminated bamboo lumber especially those made of bamboo strips have been carried out by many researchers [2] [3] [4] [5] [6] [7] [8] [9] [10]. Bamboo strips as composite product element are produced by splitting bamboo culms which have big in diameter and relatively thick wall. The bamboo culms used to produce bamboo strips as raw material for laminated bamboo lumber are commonly taken from the bottom and middle parts while the top parts are left unused, and therefore resulted in a lot of bamboo waste. To optimize bamboo utilization, the top part of bamboo culm (bamboo waste) was used as raw material for manufacture composite product other than laminated bamboo lumber such as strandboard since the element has smaller in dimension. This paper describes the results of an experiment to determine the effect of strand length and specific pressure on the properties of bamboo strandboard glued with phenol formaldehyde adhesive.

## 2. Materials and Methods

### *2.1. Materials*
The Andong bamboo *(Gigantochloa pseudoarundinacea)* culms used in the experiment were only the top parts. Twenty bamboo culms with approximately 2 m in length were cross cut into 30 – 40 cm bolts. The total bamboo bolts collected were 80. The average diameter *of G. pseudoarundinacea* was 7.25 cm while average wall thickness was 0.57 cm. Liquid phenol formaldehyde was the adhesive used in the production of the strandboards.

### *2.2. Preparation of bamboo strands*
Each bamboo bolt (30 - 40 cm in length) was manually splited green into the bamboo strips. The width of bamboo strips was approximately 2 cm. Strands were manually produced using a sharp knife with the targeted length (75 mm, 100 mm, and 150 mm), width of 2 cm and thickness of 0.5 – 0.8 mm. Only the inner part of bamboo strips were converted into bamboo strands. The bamboo strands produced were then dried in an oven until about 4% moisture content.

### *2.3. Strandboard fabrication*
Type of strandboard produced was a homogeneous uni-directionally orientation bamboo strandboard, where three different strand lengths were separately used as its composing matters. The strandboard dimensions prepared in this study was 30 cm by 30 cm by 1.2 cm. The amount of liquid phenol formaldehyde (PF) adhesive used was 7% (resin solid) based on the dry weight of strands. Based on the PF weight, 1% of para formaldehyde was used as hardener and 0.5% of wax emulsion based on the dry weight of strands was added to PF. The mixture of PF glue, para formaldehyde and wax emulsion was intensely agitated until becoming homogeneous. A specified quantity of strands was sprayed with certain amount of glue mixture in a blender. The mixture was then hand-formed into a loos mat in a wooden deckle box (30 x 30 cm internal dimensions). The wooden deckle box was removed and iron sticks (12 mm thick) were placed on the sides of the mat. The mat then loaded in a hot press set at temperature of 150°C and specific pressure of (25 kg/cm$^2$ or 30 kg/cm$^2$) for six minutes. Targeted density of strandboard produced was 0.75 g/cm$^3$. Three replications for each type of board were manufactured. The strandboards produced were conditioned for about 2 weeks before testing.

### *2.4. Testing*
Each board was cut to yield the specimens required to determine the board properties. The tests were performed using Japanese Industrial Standard (JIS A 5908) [11] for evaluating the properties of bamboo strandboards.

### *2.5. Data analysis*
A completely randomized design with factorial experiment was used in which the strand length and the specific pressure applied as the treatment factors. Three replications were prepared for each treatment combination.

IOP Conf. Series: Materials Science and Engineering **593** (2019) 012002 doi:10.1088/1757-899X/593/1/012002

## 3. Results and Discussion

### *3.1. Physical properties of bamboo strandboard*

The mean values of physical properties of strandboards are presented in Table 1. The data on physical properties of bamboo strandboard were subjected to analysis of variance (ANOVA) and the results are presented in Table 2.

The moisture content of bamboo strandboard varied from 9.0% to 9.7% with an average of 9.35%. These values are still within the range of air-dried moisture content and meet the Japanese Standard requirement for particleboard since the values are below 13%. The density of bamboo strandboard produced varied from 0.744 g/cm$^3$ to 0.772 g/cm$^3$ with an average of 0.760 g/cm$^3$. According to Dransfield and Widjaja [12] the specific gravity of *Gigantochloa pseudoarundinacea* is 0.5-0.7 (internote) and 0.6-0.8 (nodes). The bamboo strands used in the experiment were only the inner part (without the outer layers) of bamboo culms. The average density of andong bamboo strands used in this experiment was 0.56 g/cm$^3$. Therefore the compression ratio of bamboo strandboard varied from 1.33 to 1.38 with an average of 1.36.

**Table 1.** Physical properties of bamboo strandboard.

| Properties | Specific pressure, kg/cm$^2$ | Strand length | | |
|---|---|---|---|---|
| | | 75 mm | 100 mm | 150 mm |
| Moisture content, % | 25 | 9.7 (0.25) | 9.0 (0.08) | 9.5 (0.13) |
| | 30 | 9.0 (0.12) | 9.2 (0.07) | 9.7 (0.06) |
| Density, g/cm$^3$ | 25 | 0.769 (0.02) | 0.766 (0.03) | 0.754 (0.01) |
| | 30 | 0.744 (0.01) | 0.772 (0.01) | 0.755 (0.01) |
| Thickness swelling, % | 25 | 3.26 (0.20) | 2.57 (0.21) | 2.75 (0.36) |
| | 30 | 4.51 (0.35) | 3.08 (0.21) | 2.69 (0.40) |
| Wide expansion, % | 25 | 1.49 (0.02) | 1.59 (0.03) | 1.39 (0.08) |
| | 30 | 1.65 (0.13) | 1.56 (0.08) | 1.70 (0.02) |
| Water absorption, % | 25 | 25.61 (0.37) | 25.33 (1.08) | 22.88 (1.64) |
| | 30 | 24.52 (0.44) | 29.55 (0.49) | 22.75 (1.72) |

Each value was the average of three speciments.
Values in parentheses are standard deviations.

According to Maloney [13] to achieve high properties in medium density particleboard, the minimum compression ratio should be 1.3. Compression ratio is defined as the ratio of the density of the board to the density of the raw material [13]. The higher compression ratio resulted in sufficient interparticle contact area which is developed during the pressing operation to achieve good bonding. The results of ANOVA in Table 2 showed that the density of strandboard was not affected by both strand length and specific pressure.

**Table 2.** Summarized ANOVA of physical properties of bamboo strandboard.

| Properties | F calculated | | |
|---|---|---|---|
| | Strand length (A) | Specific pressure (B) | Interaction (AB) |
| Density | 1.67 $^{ns}$ | 0.72$^{ns}$ | 1.85 $^{ns}$ |
| Thickness swelling | 27.46** | 15.88** | 7.08** |
| Wide expansion | 0.26 $^{ns}$ | 18.86** | 8.15** |
| Water absorption | 26.02** | 3.64$^{ns}$ | 9.75** |

$^{ns}$ not significant
** highly significant

Thickness swelling is one indicator of dimensional stability and perform as an important property of bamboo strandboard. The thickness swelling of bamboo strandboard varied from 2.57% to 4.51% with an average of 3.14%. If the data on thickness swelling in Table 1 are compared with the Japanese Industrial Standard for Particleboard (JIS A 5908) [11] it can be seen that all bamboo strandboards produced meet the standard requirement since the values are less than 12% . The results of ANOVA in Table 2 showed that thickness swelling of strandboard was significantly affected by strand length and specific pressure. It was observed that the values of thickness swelling decreased with increasing strand length but increased with increasing specific pressure.

Hiziroglu [14] studied the properties of strandboard panels manufactured from Eastern Redcedar. The strandboard panels were produced using 8% liquid phenol formaldehyde adhesive based on oven-dry weight of strands. The results showed that thickness swelling of strandboard were 15.24 % for strandboard with the density of 0.65 g/cm$^3$ and 18.42% for strandboard with the density of 0.78 g/cm$^3$. Febrianto *et al.* [15] reported that thickness swelling of oriented strandboard prepared from steam-treated bamboo strand glued with methylene diphenyl diisocyanate (MDI, Type H3M) varied from 8.65% to 14.14%. Based on that information it can be concluded that the bamboo strandboards produced in this study are more stable than that of other composite products.

Dimensional stability of bamboo strandboard can also be measured from its wide expansion property. The values varied from 1.39% to 1.70% with an average of 1.56%. Analysis of variance was also carried out on wide expansion data. Table 2 shows that wide expansion of bamboo strandboard is only affected by strand length. Nevertheless the strong interaction between strand length and specific pressure can be found in Table 2, therefore it was seen that one can not generalize about the strand length or specific pressure effect to wide expansion property without taking into account its interaction.

Water absorption of bamboo strandboards varied from 22.9% to 29.6% with an average of 25.1%. The results of ANOVA in Table 2 showed that water absorption values was significantly affected by the strand length. It was observed that water absorption of bamboo strandboards decreased as the strand length increased. The water absorption of strandboard made from moso bamboo (*Phyllostachys pubescens*) glued with liquid diphenylmethane diisocyanate (MDI) with 6% resin content were 23.61% for board density of 0.73 g/cm$^3$ and 39.34% for board density of 0.65 g/cm$^3$ [16]. Other researchers [17] reported that water absorption of oriented strandboard made from mixture of wood strands comprising 60% *Calophyllum inophyllum* and 40% *Paulownia coreana* glued with 4.5% phenol formaldehyde was 47.2%.

*3.2 Mechanical properties of bamboo strandboard*

The mean values of mechanical properties of bamboo strandboards are presented in Table 3. The data on mechanical properties of bamboo strandboard were subjected to analysis of variance (ANOVA) and the results are presented in Table 4.

The MOR of bamboo strandboards (Table 3) varied from 564.9 kgf/cm$^2$ (55.4 MPa) to 962.3 kgf/cm$^2$ (94.4 MPa) with an average of 760.5 kgf/cm$^2$ (74.6 MPa). These values meet the Japanese Industrial Standard (JIS A 5908) [11] requirement for particleboard. The results of ANOVA in Table 4 showed that the MOR of bamboo strandboard was significantly affected by strand length. This finding was in agreement with the previous finding [18] [19] [20]. With the MOR data in Table 3 it was observed that the bamboo strandboards made of 100 mm strand-length produced the highest MOR whereas those made of 75 mm strand-length produced the lowest MOR and the strandboard made of 150 mm strand-length had MOR value higher than that of 75 mm strand-length but lower than that of 100 mm strand-length. This indicates that the distribution of strands at a 100 mm strand-length was more uniform and resulted in more sufficient interparticle contact area. Study conducted by Laemlaksakul [21] showed that the MOR value of particleboard from bamboo waste of *Dendrocalamus asper* Backer (bamboo chips) depends on density of particleboard.

**Table 3.** Mechanical properties of bamboo strandboard.

| Properties | Specific pressure, $kg/cm^2$ | Strand length | | |
|---|---|---|---|---|
| | | 75 mm | 100 mm | 150 mm |
| MOR, $kgf/cm^2$ | 25 | 564.9 (76) | 955.2 (108) | 731.3 (117) |
| | 30 | 601.2 (55) | 962.3 (49) | 748.1 (59) |
| MOE, 1000 $kgf/cm^2$ | 25 | 99.0 (10.2) | 100.7 (9.9) | 67.5 (5.3) |
| | 30 | 80.8 (4.3) | 99.3 (6.6) | 106.2 (9.4) |
| Internal bond, $kgf/cm^2$ | 25 | 5.13 (0.47) | 4.62 (0.34) | 5.58 (0.64) |
| | 30 | 6.35 (0.75) | 4.88 (0.31) | 7.22 (0.70) |
| Screw holding power, kgf | 25 | 53.9 (5.4) | 64.8 (6.1) | 51.8 (5.0) |
| | 30 | 53.6 (6.6) | 56.6 (3.7) | 58.2 (3.1) |

Each value was the average of three speciments.
Values in parentheses are standard deviations.

**Table 4.** Summarized ANOVA of mechanical properties of bamboo strandboard.

| Properties | F calculated | | |
|---|---|---|---|
| | Strand length (A) | Specific pressure (B) | Interaction (AB) |
| MOR | 31.91** | $0.27^{ns}$ | $0.05^{ns}$ |
| MOE | $4.48^{*}$ | $2.83^{ns}$ | 20.07** |
| Internal bond | 13.13** | 15.37** | $2.36^{ns}$ |
| Screw holding power | $3.14^{ns}$ | $0.08^{ns}$ | $3.05^{ns}$ |

$^{ns}$ not significant
$^{*}$ significant
$^{**}$ highly significant

MOE values of bamboo strandboards varied from 67.5 $10^3$ $kgf/cm^2$ or 6,626 MPa to 106.2 $10^3$ $kgf/cm^2$ or 10,418 MPa with an average of 92.3 $10^3$ $kgf/cm^2$ or 9,055 MPa. The bamboo strandboards produced in this study had MOE values meet the Japanese Industrial Standard requirement for type 24-10 particleboard. ANOVA in Table 4 showed that the MOE of bamboo strandboard was not only affected by the strand length but also affected by the interaction between strand length and specific pressure. Therefore one can not generalize about the strand length effect without taking into account its interaction. Previous study [19] showed that uni-directionally orientation bamboo strandboard of moso bamboo with 80 mm strand-length glued with 6% resin content of MDI had MOE value of 10.5 GPa, whereas those oriented strandboards had MOE value of 9.6 GPa. Other investigators [17] reported that MOE of oriented strandboard made from mixture of wood strands comprising 60% *Calophyllum inophyllum* and 40% *Paulownia coreana* glued with 4.5% phenol formaldehyde was 3.9 GPa. Febrianto *et al.* [15] reported that oriented strand board prepared from steam-treated bamboo strands bonded with commercial MDI adhesive had MOE value ranged from 4,828 to 10,215 MPa. Other study conducted by Laemlaksakul [21] reported that MOE of bamboo particleboard made from bamboo waste of *Dendrocalamus asper* backer (bamboo chips) glued with 13% UF was affected by density of particleboard. This finding was in agreement with the previous finding [14].

The bonding quality of any composite products is very important. To determine the bonding quality of bamboo strandboard, the internal bond strength test was carried out. The internal bond strength (IB) of bamboo strandboards varied from 4.62 $kgf/cm^2$ or 0.45 MPa to 7.22 $kgf/cm^2$ or 0.71 MPa with an average of 5.63 $kg/cm^2$ or 0.55 MPa. The data on IB were subjected to analysis of variance. The result showed that the IB of bamboo strandboard was strongly affected by strand length and specific pressure (Table 4). The lowest value of IB was obtained from bamboo strandboard produced from 100 mm strand-length and 25 kg/cm$^2$ specific pressure while the highest IB was obtained from bamboo strandboard produced from 150 mm strand-length and 30 kg/cm$^2$ specific pressure. The possible reason being that the bamboo strandboard with longer strand has more uniform distribution through the thickness of the boards hence resulting in higher IB. The bamboo strandboards produced in this study had IB values meet the Japanese Industrial Standard requirement for type 24-10 particleboard since the values are not less than 0.3 MPa. The IB of bamboo strandboard (UNID) made of 80 mm strand-length of moso bamboo and glued with 6% MDI was 0.48 MPa [19]. Other researcher [21] reported the IB of particleboard from bamboo waste of *Dendrocalamus asper* Backer (bamboo chips) bonded with 13% UF resin were 0.19MPa, 0.20 MPa, and 0.17 MPa for particleboard with the density of 600 kg/m$^3$, 700 kg/m$^3$, and 800 kg/m$^3$, respectively.

The screw holding power (SHP) of bamboo strandboards varied from 51.8 kgf or 508 N to 64.8 kgf or 635 N with an average of 56.5 kgf or 554 N. The bamboo strandboards produced in this study had SHP values meet the Japanese Industrial Standard requirement for type 24-10 particleboard since the values are not less than 500 N. The results of ANVA (Table 4) showed that the SHP of bamboo strandboards were not affected by both strand length and specific pressure. Nevertheless, it was observed in Table 3 that the SHP increased with increase in strand length. The possible reason being that the bamboo strandboard with longer strand has more uniform distribution through the thickness of the boards hence resulting in higher SHP.

## 4. Conclusion

The study concluded that Andong bamboo (*Gigantochloa pseudoarundinacea*) strands could be used as raw material for strandboard production. Strand length and specific pressure showed significant effect on some physical and mechanical properties of bamboo strandboard. The results of the current study indicated that the produced strandboards made from Andong bamboo strands glued with 7% phenol formaldehyde adhesive at various treatment combinations meet the Japanese Industrial Standard requirement for type 24-10 particleboard.

## References

[1] Widjaja E A 2012 *The utilization of bamboo: At present and for the future*. in Gintings et al. (Editors). In *Proceedings of International Seminar Strategies and Challenges on Bamboo and Potential Non Timber Forest Products (NTFP) Management and Utilization* pp 79–85.

[2] Lee A W C and Liu Y 2003 Selected Physical Properties of Commercial Bamboo Flooring. *Forest Products Journal* **53**(6) pp 23–26.

[3] Sulastiningsih I M, Nurwati and Santoso A 2005 Pengaruh Lapisan Kayu terhadap Sifat Bambu Lamina. *Jurnal Penelitian Hasil Hutan* **23**(1) pp 15–22.

[4] Rittironk S and Elnieiri M 2008 *Investigating Laminated Bamboo Lumber as an Alternate to Wood Lumber in Residential Construction in the United States* (Changsa, China: Proceedings of First International Conference on Modern Bamboo Structures (ICBS-2007)) October, 28–30 pp 83–96

[5] Correal J and Lopez L 2008 *Mechanical Properties of Colombian Glued Laminated Bamboo in Xiao et al (Editors). Modern Bamboo Structures.* (Changsa, China: Proceedings of First International Conference on Modern Bamboo Structures (ICBS-2007)) October, 28–30 pp 121–127

[6] Sulastiningsih I M and Nurwati 2009 Physical and Mechanical Properties of Laminated Bamboo Board *J. Trop. For. Sci.* **21**(3) pp 246-251

[7] Correal J and Ramirez F 2010 Adhesive Bond Performance in Glue Line Shear and Bending for Glued Laminated Guadua Bamboo. *J. Trop. For. Sci.* **22**(4) 433–439

IOP Conf. Series: Materials Science and Engineering **593** (2019) 012002 doi:10.1088/1757-899X/593/1/012002

[8] Mahdavi M, Clouston P L and Arwade S R 2011 Development of Laminated Bamboo Lumber: Review of Processing, Performance, and Economical Considerations. *Journal of Materials in Civil Engineering* **23**(7) pp 1036–42
[9] Sulastiningsih I M and Santoso A 2012 Pengaruh Jenis Bambu, Waktu Kempa dan Perlakuan Pendahuluan Bilah Bambu terhadap Sifat Papan Bambu Lamina. *Jurnal Penelitian Hasil Hutan* **30**(3) pp 198–206
[10] Sulastiningsih I M, Santoso A and Krisdianto 2016 Karakteristik balok bambu lamina susun tegak dari bambu Andong (*Gigantochloa pseudoarundinacea* (Steud.) Widjaja). *Jurnal Penelitian Hasil Hutan* **34**(3) pp 167–177
[11] Japanese Industrial Standard 2016 *Particleboards* (JIS A 5908:2015). (Tokyo: Japanese Standards Association)
[12] Dransfield S and Widjaya E A 1995 *Plant Resources of South East Asia No 7 Bamboos*. (Leiden: Backhuys Publishers)
[13] Maloney T M 1993 *Modern Particleboard and Dry Process Fibreboard Manufacturing*. (San Fransisco: Miller Freeman Publications, Inc)
[14] Hiziroglu S 2009 Properties of Strandboard Panels Manufactured from Eastern Redcedar. *Materials* **2** 926-933
[15] Febrianto F, Jang J-H, Lee S-H, Santoso I A, Hidayat W, Kwon J-H., and Kim N-H 2015 Effect of Bamboo Species and Resin Content on Properties of Oriented Strand Board Prepared from Steam-treated Bamboo Strands. *BioResources* **10**(2) pp 2642–55.
[16] Sumardi I and Suzuki S 2014 Dimensional stability and mechanical properties of strand-board made from bamboo. *BioResources* **9**(1) pp 1159–67
[17] Oh Y S and Kim J M 2015 Properties of oriented strandboard bonded with phenol-urea-formaldehyde resin. *J Trop For Sci* **27**(2) pp 22–26
[18] Febrianto F, Sahroni, Hidayat W, Bakar E S, Kwon G-J, Kwon J-H, Hong S-I and Kim N-H 2012 Properties of oriented strand board made from Betung bamboo (*Dendrocalamus asper* (Schultes.f) Backer ex heyne). *Wood Sci Technol* **46** pp 53–62
[19] Sumardi I, Suzuki S and Rahmawati N 2015 Effect of board type on some properties of bamboo strandboard. *J Math Fund Sci* **47**(1) pp 51–59
[20] Sulastiningsih I M, Indrawan D A, Balfas J, Santoso A and Iskandar M I 2017 Sifat fisis dan mekanis papan untai berarah dari bamboo tali (Gigantochloa apus (J.A. & J.H. Schultes) Kurz). *Jurnal Penelitian Hasil Hutan* **35**(3) pp 197–209
[21] Laemlaksakul V 2010 Physical and mechanical properties of particleboard from bamboo waste. World Academy of Science, Engineering and Technology *International Journal of Chemical, Moleculer, Nuclear, Materials and Metallurgical Engineering* **4** pp 4

IOP Conf. Series: Materials Science and Engineering **593** (2019) 012003 doi:10.1088/1757-899X/593/1/012003

# Relationship between thermal conductivity and adhesive distribution of phenol-formaldehyde visualized with x-ray computed tomography on sugi (*Cryptomeria japonica* D.Don) heartwood plywood

**N Ferrtikasari[1*], T Tanaka[1], M Yamada[1]**

Department of Agriculture, Graduate school of Integrated Science and Technology, Shizuoka University, Ohya 836, Suruga-ku, Shizuoka City 422-8529 Japan

*Email: nindya.ferrtikasari.16@shizuoka.ac.jp

**Abstract**. The relationship of transverse thermal conductivity of non-resin region and resin-rich region in softwood plywood with various relative resin thicknesses in lumen of softwood has been determined theoritically. Futhermore, Interest in influence of adhesive penetration on adhesive bond performance and adhesive behavior has been evident by using microscopy technique and X-ray apparatus for several decades. Therefore, the relationship between thermal conductivity and adhesive distribution on softwood plywood product is needed to reveal as the development of material properties by visualizing X-ray CT apparatus. Micro Focus X-ray Computed Tomography (X-ray CT) was used to analyse the adhesive distribution of sugi heartwood plywood for different amount of glue spreading rates which were applied on the veneer surface. Four plywood samples (5-ply) were produced by spreading the phenol-formaldehyde resin adhesive on the veneer surface (dimension 450 mm× 450 mm× 3.81 mm) as much 0, 75, 150, and 225 g/m$^2$ of glue spreading rates. All the veneer samples were conditioned 30 days under 20$^o$C and 65% RH before plywood manufacturing. The veneer samples were dried at 60 temperature 60$^o$C for 24 hours before pressing by cold press apparatus, the plywood samples were pressed (1 MPa) for 20 minutes and hot press apparatus at 125$^o$C for 19 minutes. The analysis of thermal conductivity apparatus followed standards are JIS A 1412 and ASTM- C518. The plywood samples (dimension 200 mm× 200 mm ×thickness mm) in air dried condition were put on among plates of thermal conductivity apparatus for 40 minutes for each sample to obtain the thermal conductivity values. The results seemed to indicate that the thermal conductivity increases with increasing amount of glue spreading rates. By visualizing with the X-ray CT was revealed that the largest amount of glue spreading rate (225 g/m$^2$) had the highest distribution of PF adhesive penetrated within the lathe checks of veneer sheet compared to 150, 75, and 0 g/m$^2$. Furthermore, increasing of the adhesive distribution into the lathe checks of veneer increased the plywood density. According to the results, we suggested that the X-ray CT images analysis was powerful to reveal the relationship between the thermal conductivity and the adhesive distribution.

## 1. Introduction

Japanese cedar (*Cryptomeria japonica* D. Don) is a popular commercial wood for softwood plywood manufacturing in Japan. In Japanese composite panel market, the production of domestic softwood plywood has been increasing rapidly, and plywood industry has started to peel sugi logs to produce thicker panels [1]. Thermal conductivity of wood composite materials such as plywood, OSB, particleboard, and fiberboard has varied according to wood species, direction of wood fiber, ratio of early and late wood, thickness of composite materials, density, moisture content, resin type and additives, temperature, and flow direction of heat in several studies [2]. Therefore, the adhesion of wood adhesive in wood-working industry could be an important factor of the quality parameters of plywood which affects bonding strength, permeability, heat and moisture transfer, etc.

Controlling the amount of glue spreading rate on the surfaces of veneer in plywood production is important since it will affect the final products properties, manufacturing costs and the environmental impact as well [3]. In plywood production, human operators find it difficult to precisely monitor the spread rate of adhesive in real-time [4]. According to Lindgren (1992), Kamke and Lee (2007), and Paris and Kamke (2015), the X-ray CT had visualized the properties of wood and appearance of the adhesive distribution performance inside the wood-based panel's products [5, 6, 7]. They visualized the adhesive distribution performance for improving the technique in composite science scope. The objective of this study was to elucidate the effect of glue spreading rate of plywood on its thermal conductivity and to investigate the mechanism of the effect by visualizing the adhesive morphology across glue lines of plywood by using a micro focus X-ray CT imaging.

## 2. Materials and Methods

### *2.1. Materials*

#### *2.1.1. Veneer sheets*

The plywood samples were produced made by veneer sheets of hearwood of the Japanese cedar or Sugi (Cryptomeria japonica. D.Don) grown in Shizuoka Prefecture, Japan. Twenty heartwood veneer sheets with the dimension average was 450 mm × 450 mm × 3.81 mm (thickness) made from Japanese cedar were prepared and conditioned at 20°C and 65% RH at least for 30 days. The dimension and weight of each sheet was measured and its density also was calculated (Table 1).

*2.1.2. Phenol-formaldehyde resin adhesive.* Phenol-formaldehyde resin adhesive was prepared with the following ratio: PF resin = 76.3%, $CaCO_3$ = 11.6%, Wheat powder = 6%, Distilled water = 3.8% and $Na_2CO_3$ = 2.3%.

### *2.2. Methods*

#### *2.2.1. Plywood Manufacturing*

The veneer sheets were dried at temperature 60°C for 24 hours in a drying oven prior to plywood manufacturing. Four specimens of 5-ply plywood were made with the veneer sheets with the glue spreading rate of 0 (no adhesive), 75, 150, and 225 g/cm$^2$ (single glue line) with 20-minutes cold press at room temperature (1 MPa) followed by hot press at 125°C for 15–19 minutes (1 MPa). The samples were then conditioned at least for 30 days and had air dried condition before testing of the thermal conductivity (Table 1).

**Table 1.** The properties of plywood materials.

| No sample | Veneer density of sugi heartwood (g/cm$^3$) | Spread rate (g/cm$^2$) | Plywood density (g/cm$^3$) | Plywood thickness (mm) |
|---|---|---|---|---|
| 1 | 0.319, 0.334, 0.335, 0.341, 0.343 | 0 | 0.366 | 18.88 |
| 2 | 0.321, 0.334, 0.345, 0.335, 0.340 | 75 | 0.397 | 17.68 |
| 3 | 0.344, 0.336, 0.339, 0.323, 0.333 | 150 | 0.425 | 17.04 |
| 4 | 0.346, 0.325, 0.337, 0.331, 0.338 | 225 | 0.448 | 16.27 |

*2.2.2. Thermal conductivity measurement*

The plywood specimens were cut into 20 × 20 cm in the air dried conditions (10-12 % MC). The thermal conductivity of each specimen in the thickness direction was measured with a heat flow method between two plates set at 10 and 40°C (AUTO-Λ HC073, Eko Instrument Trading Co.LTD). The thermal conductivity apparatus was following standard are JIS A 1412 and ASTM- C518 (Figure 1).

**Figure 1.** Thermal conductivity measurement apparatus.

*2.2.3. X-ray CT Scanning*

The X-ray CT images of the four specimens were taken with a micro focus X-ray CT scanning system (Inspexio SMX-225 CT, Shimadzu Corporation) with 80 kV tube voltage and 150μA tube current (Figure 2).

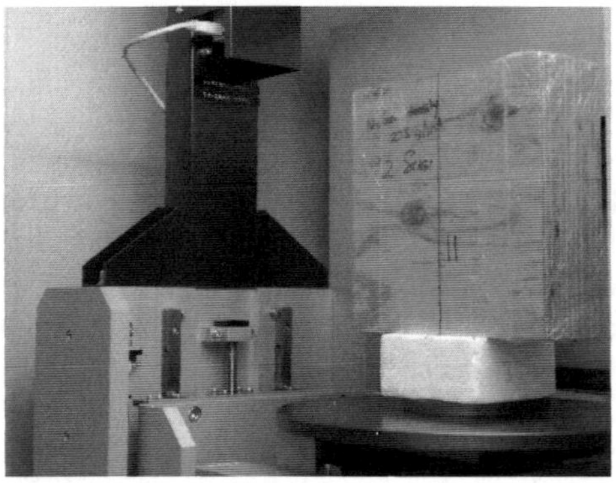

**Figure 2.** A micro focus X-ray CT system.

## 3. Results and discussion

Figure 3 shows the thermal conductivity increases with increasing the amount of glue spreading rate of sugi heartwood plywood. The big amount of glue spreading rate 225 g/m$^2$ had the highest thermal conductivity value, followed by plywood had amount of glue spreading rate were 150, 75 and 0 g/m$^2$. As shown in figure 3 that the highest plywood density had the largest thermal conductivity, which agrees with results presented by Sonderegger and Niemz (2009) and Tanaka et al, (2012) [8, 9]. The density of plywood was influenced by the amount of glue spreading rates, which were spread on the veneer surface (Figure 3). The biggest amount in application of resin adhesive had the biggest adhesive distribution are spread wider and deeper within the plywood (Figure 5). Increasing of the adhesive distribution into the lathe checks of wood increased the plywood density. There are seems small effects with increasing of amount of spreading rate on thermal conductivity.

Figure 4 shows the differences of adhesive distribution appearance by using micro focus X-ray CT vary in amount of glue spreading rates (0, 75, 150 and 225 g/m$^2$). By visualizing of the X-ray CT investigated that the largest amount of glue spreading rate (225 g/m$^2$) (Figure 5) had the highest distribution of PF-adhesive penetrated within the lathe checks of veneer sheet compared to 150, 75, and 0 g/m$^2$ (Figure 6, 7 and 8). Figure 5 showed that the transverse view of veneers bright part is distribution of resin in glue lines and lathe checks of veneer dominantly and black parts are veneers. Rohumaa et al., (2013) studied the influence of the lathe check depth and orientation on bond quality of phenol-formaldehyde bonded birch plywood [10]. The study had revealed that by the deeper checks allowing the adhesive penetrates much more deeply into the veneer which makes strong influence on the measured shear strength. As depicted in Figure 6 and 7, the adhesive distribution performances (150 and 75 g/m$^2$) were found penetrating on the glue lines and lathe checks very slight which affected thermal conductivity values smaller than 225 g/m$^2$. As visible in figure 8, the plywood with 0 g/m$^2$ amount of glue spreading rate had the smallest thermal conductivity. The result was supported by visualizing images that there were air contact lines depicted as black parts. The air contact lines on among veneers make the plywood could not bond tightly when thermal conducted transversely. According to the results, Interest into influence of adhesive penetration on adhesive bond performance and adhesive behavior of different wood-adhesive glue lines has been evident by using microscopy technique and X-ray apparatus for several decades [6,7]. Therefore, we suggested that the X-ray CT images analysis was powerful to reveal the relationship between the thermal conductivity and the adhesive distribution.

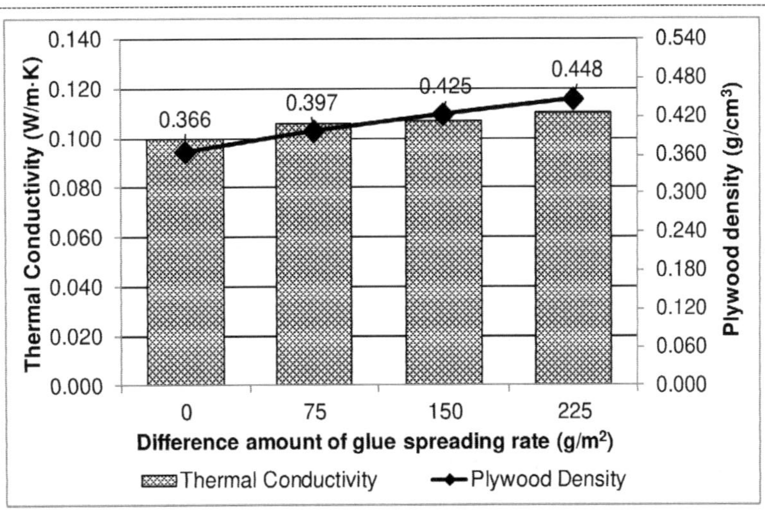


**Figure 3.** Graphic of relationship between adhesive distribution and thermal conductivity with plywood density's consideration.

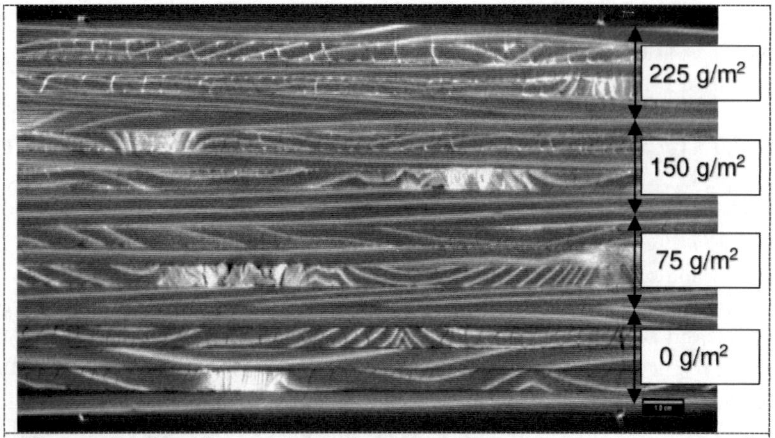


**Figure 4.** Micro Focus X-ray Computed Tomography data image of PF-adhesive distribution within sugi plywood.

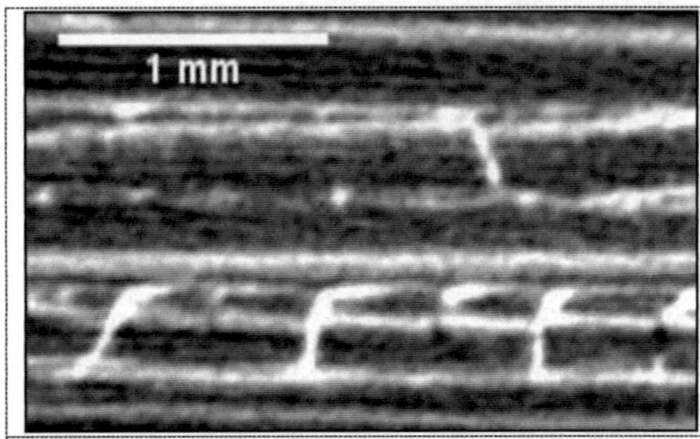


**Figure 5.** Micro Focus X-ray Computed Tomography (X-ray CT) image of PF- adhesive distribution (225 g/m$^2$) in sugi plywood. The transverse view of veneers bright parts are distribution of resin in glue lines and lathe checks of veneer dominantly and black (dark) parts are veneer sheets.

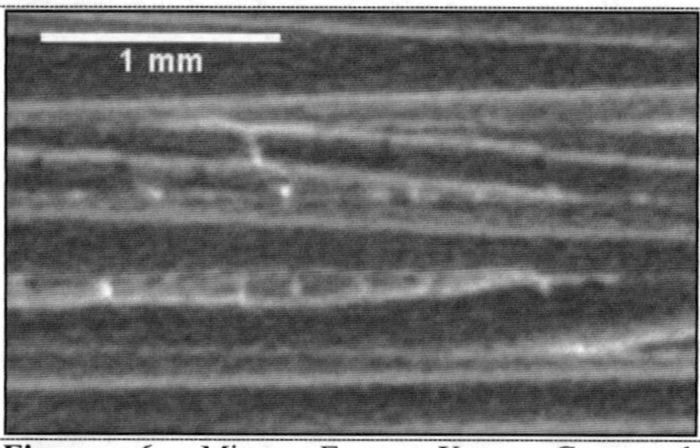


**Figure 6.** Micro Focus X-ray Computed Tomography (X-ray CT) image of PF- adhesive distribution (150 g/m$^2$) in sugi plywood. The transverse view of veneers bright parts are distribution of resin in glue lines and lathe checks of veneer also appearance of the annual growth rings of wood.

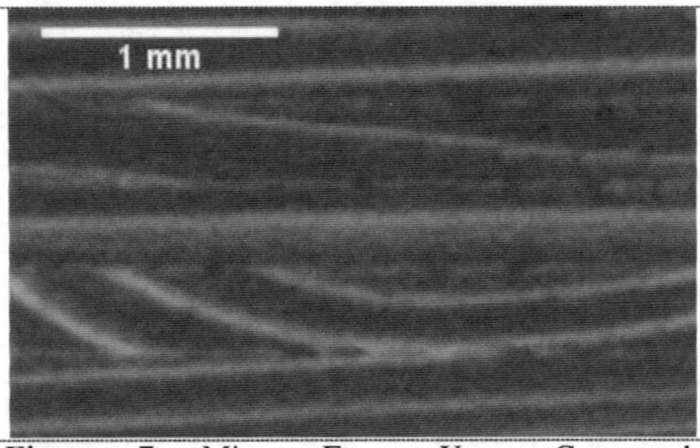


**Figure 7.** Micro Focus X-ray Computed Tomography (X-ray CT) image of PF- adhesive distribution (75 g/m$^2$) in sugi plywood. The adhesive distribution is slight to visible by X-ray CT of glue spreading rate is 75 g/m$^2$. The view of veneers bright parts are visualized of the annual growth rings of sugi.

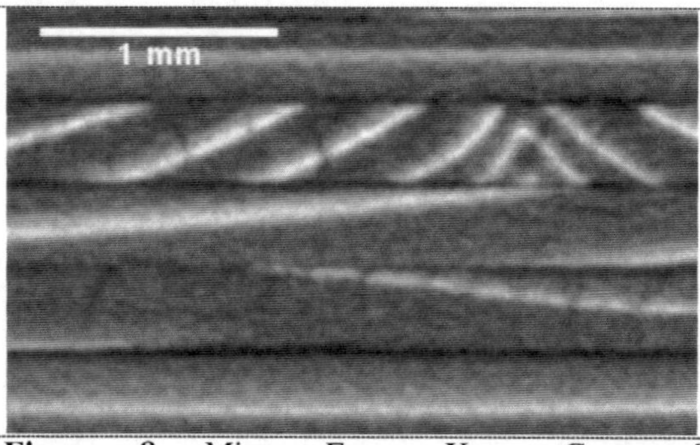


**Figure 8.** Micro Focus X-ray Computed Tomography (X-ray CT) image of the plywood sample without the application of adhesive in sugi plywood. The view of veneers bright part are visualized of growth annual rings of sugi and black (dark) views are the air contact lines.

## 4. Conclusions

Based on the results, the following conclusions could be concluded:

- Thermal conductivity of sugi heartwood plywood increases with increasing the amount of glue spreading rate by using phenol-formaldehyde resin adhesive.
- X-ray CT images analysis are powerful to reveal the relationship between the thermal conductivity and the adhesive distribution.

## References

[1] Suzuki, S 2005 *Using Cedar Plantation materials for wood-based composite in Japan. In proceedings of Scientific Session 90 "Using Wood Composite as a tool for Sustainable Forestry"* (Brisbane, Australia: XXII IUFRO World Congress) August 12 pp. 9–13

[2] Demirkir C, Colakoglu G, Colak S, Aydin I, and Candan Z 2016 Influence of aging procedure on bonding strength and thermal conductivity of plywood panels *Acta Physica Polonica A* **129** (6) pp 1230–1234

[3] Marra A A 1992 *Technology of wood bonding*. (New York: Van Nostrand Rein Hold) pp 35–54.

[4] Antikainen T, Rohumaa A, Hunt C G, Levirinne M, and Hughes M 2015 Estimating the spread rate of urea formaldehyde adhesive on birch (Betula pendula Roth) veneer using fluorescence. *Eur. J. Wood Prod* **73** 69–75

[5] Lindgren O, Davis J, Wells P, and Shadbolt P 1992 Non-destructive wood density distribution measurements using computed tomography *Holz als Roh- und Werkstoff* **50** pp 295–299

[6] Kamke F A and Lee J N 2007 Adhesive penetration in wood-A review. *Wood and Fiber Sci.* **39**(2) pp 205-220

[7] Paris J L and Kamke F A 2015 Quantitative wood-adhesive penetration with X-ray computed tomography. *Int. J. Adhes. Adhes.* **61** pp 71–80

[8] Sonderegger W and Niemz P 2009 Thermal conductivity and water vapour transmission properties of wood-based materials. *Eur J wood prod* **67** (3) pp 313–321

[9] Tanaka T, Shibusawa T, Asakura, and Shida S 2012 *Moisture and thermal properties of wood-based panels (II) water vapor permeability* (Shizuoka, Japan: Proceedings BIOCOMP2012 11th Pacific Rim Bio-Bases Composite Symposium ) pp 425–433

[10] Rohumaa A, Hunt C G, Hughes M, Frihart C R and Logren J 2013 *Holzforschung* **67** (7) pp 779-786

**Acknowledgments**

The Authors express their appreciation to JSPS KAKENHI for funding (Fund No. 17H05032), NODA Corp for providing the samples of sugi veneer sheets, J-Chemical for supplying a phenol-formaldehyde resin adhesive, Dr. Hidefumi Yamauchi, Dr. Yasuo Kawai and Ms. Aoi Oishi for helping the measurement of thermal conductivity.

IOP Conf. Series: Materials Science and Engineering **593** (2019) 012004 doi:10.1088/1757-899X/593/1/012004

# Thermal Properties of Isocyanate as Particleboard's Adhesive

**A Nuryawan[1*], E M Alamsyah[2]**

[1]Department of Forest Products Technology, Faculty of Forestry, University of Sumatera Utara, Jl. Tri Dharma Ujung No.1 Kampus USU Medan, North Sumatera, Indonesia 20155
[2]School of Life Science and Technology, Bandung Institute of Technology, Jl.Ganesha No.10 Bandung, West Java, Indonesia

*Email:arif5@usu.ac.id

**Abstract.** The objective of this study was to evaluate the thermal properties of isocyanate as particleboard adhesives. In this study, thermal properties of the isocyanate was investigated particularly its curing behavior either using water or wood with a temperature range from 25 to 500°C using thermal analyzer (Mettler Toledo). When isocyanate is used as the binder in particleboard system, the reactive –NCO group reacted with water (because wood is a hygroscopic material and contains free water and bound water) and also with the –OH group from wood. Sample of isocyanate was H3M purchased from PT. Polichemie Asia Pacific and wood particle was *Eucalyptus* sp. taken from PT. Toba Pulp Lestari. About 10 mg of the mixture either isocyanate with water or wood in various weight ratios were scanned at four different heating rates (2.5; 5; 10; and 20°C/min). Results of this study showed isocyanate cured either using water or moisture wood particle indicated by the *Tp* or peak temperature on the thermograms of differential scanning calorimetry (DSC). However when isocyanate reacted with wood particle, the *Tp* emerged more than two which indicated isocyanate also reacted with wood components. Furthermore, values of enthalpy seemed irregular probably due to different structures and compositions of wood particle. This influence, however, was masked by the moisture present in the wood (free and bound water), because water was much easier to react with isocyanate than were the –OH within the wood particle. It was suggested to calculate activation energy since the heating rates used up to four and to understand whether water, wood particle, or both affected the main reaction mechanism among isocyanate, water within the wood (free and bound water), or wood components.

## 1. Introduction

A significant market was born for particleboard bonded by isocyanate in the late 1960s and early 1970 after the proven that isocyanate was effective for bonding particleboard [1]. Data from [2] strengthen that isocyanate has been placed third rank in worldwide market adhesive for wood after urea-formaldehyde (UF) and melamine-urea-formaldehyde (MUF) resin, respectively. Up to now, isocyanates have been applied for gluing in engineered wood products especially wood-laminating industry as the urethane prepolymers originated from isocyanate-polyol reaction products and as the isocyanate have been served in the particleboard industry [3]. In other words, the type of isocyanate generally used for the first is the water-based emulsion adhesives with isocyanate as the cross-linker [4] while the type of isocyanate generally used for the latter is MDI or 4-4'-diphenylmethane diisocyanate [3]. The methods of mass production for particleboard using this adhesive have been patented in Europe since 2013 [5]. A comprehensive review describes the development of these both adhesives in Indonesia has been published recently [6].

In this following study will be limited to particleboard isocyanates, since world production of particleboard reaches up to 93 million $m^3$ per year nowadays [7]. Further, some tremendous application researches have been emerged in relation with particleboard bonded isocyanate, for example Papadapolous [8] compared between conventional particleboard bonded UF resins and the one bonded isocyanate. The latter gave superior properties even though there was addition of wax around 1%. Modification UF resin with isocyanate for bonding particleboard has been done by Dziurka & Mirski [9] and Nuryawan et al. [10]. The similar results have been obtained that physical and mechanical properties of the particleboard were enhanced with the increase of isocyanate loading to the UF resin. The reason for the excellent performance of isocyanate compared to another wood adhesive is still a matter of debate. Zhou and Frazier [11] showed that isocyanate is able to form covalent urethane bonds with hydroxyl groups of wood components, which was interpreted as a contribution to strong bondline behavior. However, the major reaction of isocyanate takes place with water situated in the wood substrate to form polyurea/biuret networks [12-13]. In this respect, infra red spectroscopy was employed to investigate the evidence of the interaction between isocyanate with water or wood [14-15]. Further, because manufacturing of particleboard involved hotpressing, some thermal analytical methods have been applied for investigating the thermal properties of wood particle-isocyanate adhesive system. For example Umemura et al. [16-17] reported the durability of unique sample of isocyanate assessed by DMA (dynamic mechanical analysis). He and Yan [18] stated that curing kinetics of isocyanate have been influenced by moisture in wood as well as wood species in limited moisture therein. Recently, the authors presented thermal stability of isocyanate when exposed to water or hygroscopic wood particle assessed by TGA (thermogravimetric analysis) to investigate the resistance to heat including its degradation [19].

Excess water or mositure in the particleboard system resulted in blister or blowing particleboard. On the contrary, starve water or moisture in the particleboard resulted in less strength. In other words, wood particle-isocyanate adhesive system is exposed rapidly changing conditions of pressure, steam and heat during manufacture. Therefore, we evaluate thermal properties of isocyanate as particleboard's adhesive, including curing characteristics of isocyanate using water or moisture wood particle. The objective of this work is to study the fundamental aspect of thermal properties of the isocyanate, particularly its transformation temperature and enthalpy when isocyanate exposed to water or hygroscopic wood particle. Therefore, in this contribution, isocyanate was mixed with either water or wood particle and scanned under differential scanning calorimeter (DSC) because a little report available related to the nature of free and bound water within the wood particle.

## 2. Materials and methods

Isocyanate used was product of PT. Polichemie Asia Pacific (Medan branch, North Sumatera, Indonesia) with trademark of H3M having 98% solid content. *Eucalyptus* sp. wood particles with 20-mesh in size were taken from PT. Toba Pulp Lestari (Porsea, North Sumatera, Indonesia). For accurate analysis, measurement of moisture content was conducted using gravimetric analysis prior to mix with isocyanate, resulted in 6.20%.

### *2.1. Formulation mixture*

Simple mixing isocyanate with either water or wood particle at various proportions based on weight (w/w) was applied in order to develop the mixture. Isocyanate was placed into a beaker glass and either water (1%; 3%; 5%; and 7%) or wood particle (5%; 10%; 12%; and 15%) was added. Both components were thoroughly stirred by glass spatula about 5 minutes or until they looked physically mixed. Addition of water was much less than wood particle with considering wood contains water therein (free and bound water).

### *2.2. Differential scanning calorimetry (DSC)*

Differential scanning calorimetry (DSC) measurements were performed on TGA/DSC1 Star system Mettler Toledo, operating at Institute of Advance Materials-University Putra Malaysia (Selangor, Malaysia). For each measurement, about 10 mg of the mixture either isocyanate with water or wood in various weight ratios were scanned to a temperature gradient ranging from 25 to 500°C with four different heating rates (β) of 2.5; 5; 10 and 20°C/mins using nitrogen gas 50.0 ml/mins. The enthalpy changes were recorded and analyzed for the enthalpy integral (*ΔH*) and the peak maximum (*Tp*) using a software package of STARe 15.00. For the control, prior scanning the mixtures, each component (isocyanate/ IC and wood particle/ WP) was scanned at normal heating rate of 10°C/mins.

## 3. Results and Discussions

### *3.1. DSC measurements of the isocyanate and wood particle in normal heating rate*

Figure 1 showed typical DSC's thermograms of main components of the mixture, namely IC and WP at heating rate of 10°C/minutes.

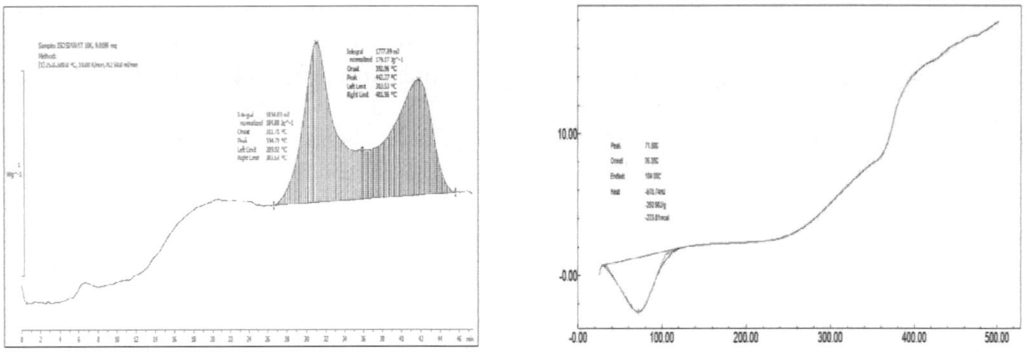

(a) (b)

**Figure 1**. Typical thermograms of DSC's isocyanate (a) and wood particle (b) at β=10°C/mins.

There were two *Tp* of the IC's thermogram (Figure 1.a.) presumably because of the difference in the molecular weight of the composition as well as the chemical species [6]. According to [20], MDI is available in three broad types of compositions, namely polymeric MDI (PMDI), monomeric MDI (MMDI), and modified MDI which have different characteristics as well, such as functionality, odor, -NCO content, viscosity, freeze point, boiling point, and density. The distinct characteristics of the compositions may result in different *Tp*. An endothermic *Tp* of the WP's thermogram (Figure 1.b) related to water vapor. Water originated from free water in the lumen and bound water within cell-walls. In fact, the moisture content of the WP was 6.20%. Detail DSC data measurements, including onset and endset temperature, *To* and *Te*, and enthalpy (*ΔH*) presented in Table 1.

**Table 1.** Comparison of DSC data of the IC and WP at β=10°C/ mins.

| Sample | *β* (°C/mins) | Temperature (°C) | | | *ΔH* (J/g) |
|---|---|---|---|---|---|
| | | *To* | *Tp* | *Te* | |
| IC | 10 | 312 | 335 | 384 | 185 |
| | | 393 | 442 | 482 | 179 |
| WP | 10 | 35 | 72 | 104 | 251 |

### *3.2. Influence of amount water and heating rate to the DSC measurement*

Figure 2 showed a representative DSC's thermograms of mixture of IC and water at two different β. Mixture of IC and water, less amount of water (Figure 2.a) or much amount of water (Figure 2.b), resulted in consistent number of *Tp*, for each measurement yielded two *Tp*.

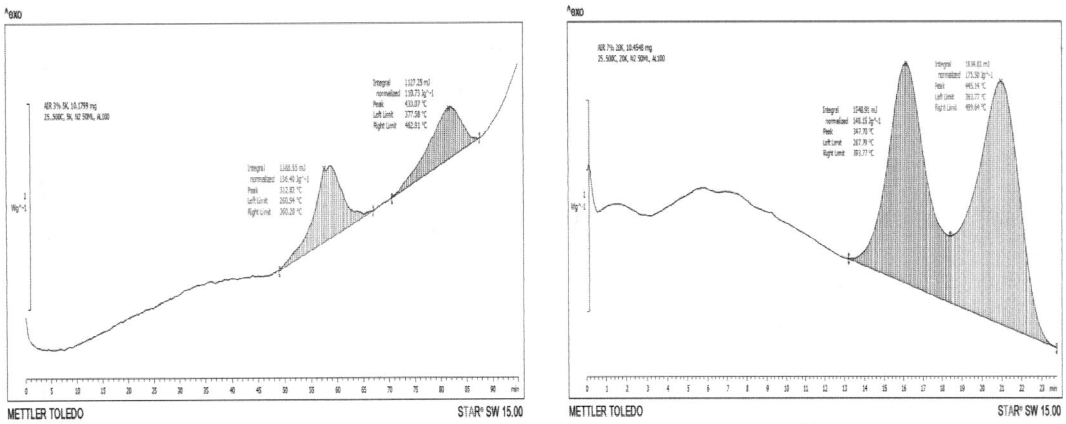


(a) (b)

**Figure 2**. Typical representative thermograms of DSC's mixture of low ratio of water (IC and 3% water (w/w) at slow β=5°C/mins (a) and high ratio of water (IC and 7% water (w/w) at fast β= 20°C/mins (b).

As shown in presented data in Table 2, *To*, *Tp*, and *Te* occurred at high temperature. IC reacts with water and the rate of the reaction is very slow at temperature below 50°C [20]. The first *Tp* around 300°C represented the early stage of curing and the second *Tp* around 400°C represented the latter stage of curing. Therefore in this present study, the early part of cure is dominated by chain extension of IC pre-polymer and little is contributed to strength development. The latter stages of cure are dominated by the development of a polymer network. During the development of the three-dimensional network adhesives strength grows as well [21]. Unfortunately the reaction of IC with water liberates $CO_2$ gas and forms insoluble polyureas as shown in the equation (1)

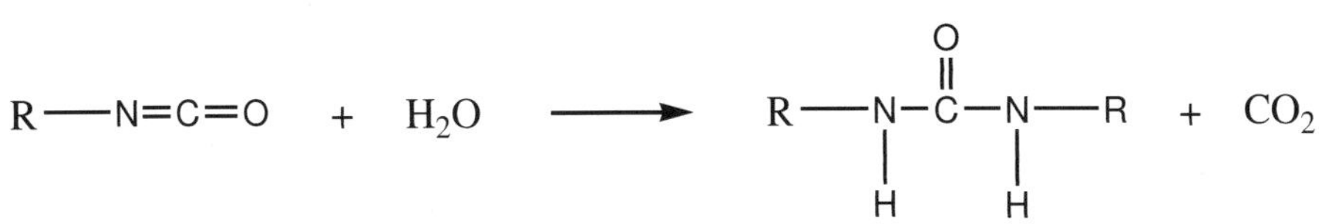

................(1)

β as the rate of temperature increase in a response to a temperature program [22] seemed affected the *Tp* in isocyanate-water system. Fast β (20°C/mins) resulted in higher *Tp* and slow β (2.5°C/mins) yielded lower *Tp* except for less amount of water (≤ 3%). This phenomenon can be explained using kinetic modelling for achieving a better understanding of the curing occurrence. In case of IC-water system, the cure reaction always involves the consumption of –NCO group. Kinetic parameters can be estimated from multiple ramps with different β at least four and the calculation will be published in upcoming publications. By monitoring the kinetics modelling of IC cured with water, the insight can be gained even though further analysis is needed to ensure the mechanism of the reaction between IC and water.

**Table 2.** Data obtained from DSC thermograms including $T_o$, $T_p$, $T_e$ and $\Delta H$ at different β from measurement of sample made of mixture IC and water at various ratios.

| Sample | β (°C/mins) | Temperature (°C) | | | $\Delta H$ (J/g) |
|---|---|---|---|---|---|
| | | $T_o$ | $T_p$ | $T_e$ | |
| IC+1% water | 2.5 | 349 | 419 | 433 | 327 |
| | | 433 | 454 | 499 | 336 |
| | 5 | 263 | 312 | 366 | 144 |

| | | | | | |
|---|---|---|---|---|---|
| | | 366 | 431 | 463 | 140 |
| | 10 | 282 | 331 | 377 | 160 |
| | | 377 | 437 | 477 | 168 |
| | 20 | 291 | 350 | 388 | 134 |
| | | 388 | 448 | 501 | 143 |
| IC+3% water | 2.5 | 350 | 412 | 438 | 204 |
| | | 438 | 451 | 498 | 185 |
| | 5 | 269 | 313 | 360 | 136 |
| | | 378 | 433 | 463 | 111 |
| | 10 | 267 | 328 | 378 | 173 |
| | | 378 | 435 | 477 | 201 |
| | 20 | 289 | 349 | 387 | 135 |
| | | 387 | 445 | 493 | 154 |
| IC+5% water | 2.5 | 261 | 305 | 345 | 281 |
| | | 345 | 420 | 499 | 849 |
| | 5 | 273 | 317 | 361 | 124 |
| | | 373 | 431 | 463 | 118 |
| | 10 | 291 | 339 | 384 | 165 |
| | | 384 | 443 | 488 | 157 |
| | 20 | 285 | 350 | 390 | 154 |
| | | 389 | 449 | 494 | 173 |
| IC+7% water | 2.5 | 259 | 289 | 353 | 115 |
| | | 363 | 413 | 442 | 124 |
| | 5 | 269 | 312 | 365 | 134 |
| | | 365 | 432 | 466 | 145 |
| | 10 | 283 | 337 | 367 | 155 |
| | | 367 | 442 | 491 | 202 |
| | 20 | 288 | 348 | 394 | 148 |
| | | 394 | 445 | 500 | 176 |

### *3.3. DSC of the mixture (isocyanate and wood particle) at various composition and different heating rates*

Figure 3 showed a representative DSC's thermogram of mixture of IC and WP at two different β. The mixture of IC and WP, with less amount of WP (Figure 3.a) or much amount of WP (Figure 3.b) using slow β and fast β, respectively. Unexpectedly, slow β (2.5 and 5°C/mins) resulted in three up to four *Tp* consistently for each measurement as shown in detail in Table 3. On the contrary, normal β (10°C/mins) and fast β (20°C/mins) resulted in only two *Tp*. This results indicated that β influenced the resulted *Tp*. Most of early heat generation prior to the maximum is likely IC consumption contributing to increase in the molecular weight of the polymer, the exothermic peak doesnot represent structures or reaction products that form late in cure. The variation in peak conversion with the β can reflect the transition of the reaction features [18].

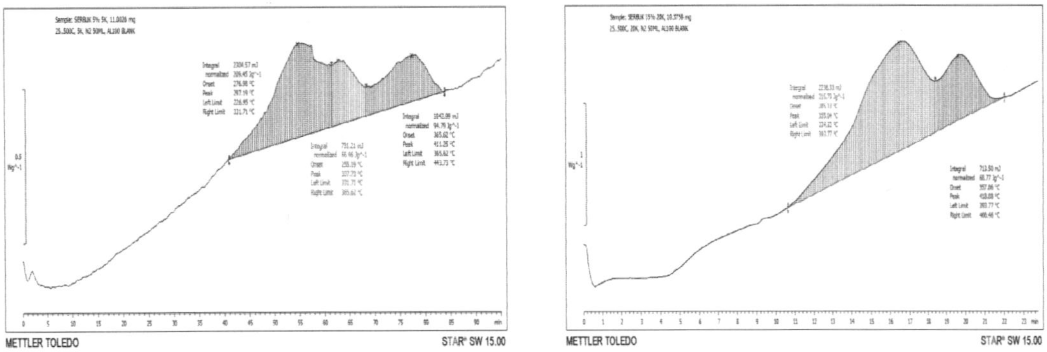


(a) (b)

**Figure 3**. Typical representative thermograms of DSC's mixture of low ratio of WP (IC and 5% WP (w/w) at slow β=2.5°C/mins (a) and high ratio of WP (IC and 15% WP (w/w) at fast β= 20°C/mins (b)

Indeed, isocyanate reacted with either free or bound water from WP but if the water is present in the wood materials, -NCO group within the IC would react with water in preference [15]. Equation (2) confirmed that -NCO group within IC reacted with -OH from wood provided set of combination and variety of polymer containing ester, ether, urea, alkyl or aryl units that affect thermal stability of the material [23].

$$R{-}N{=}C{=}O \quad + \quad HO\text{-}wood \longrightarrow R{-}N(H){-}C({=}O){-}O{-}wood \qquad .......(2)$$

There was some controversy on the various reactions during IC reacting with wood, based on the chemistry of IC, that the main reaction of IC with WP in the presence of moisture may include primary and secondary reactions. First, IC reacts with primary and secondary hydroxyl groups (-OH) within the wood to form urethane and with water absorbed in the wood to form urea. Second, urethane and urea then react with IC, as the secondary reaction to form allophanate and biuret. Studies of [24,25] on IC/wood bondline by nuclear magnetic resonance spectroscopy that the curing chemistry of IC with dried wood resulted predominantly in biuret-type structures from the IC reaction with a small amount of moisture. When the pre-cure moisture content of wood was increased to 4.5%, the end products of curing were predominantly polyurea-type structures with less biuret formation and residual isocyanate. In other words, the -OH groups in wood may react with IC to form polyurethane, providing direct covalent linkages between the adhesive and wood. However, the reaction of IC with water absorbed in wood to form polyurea and biuret-type structures can compete strongly with the urethane formation.

In this case, IC reacted with either free or bound water from WP first, then continued by wood components; whether cellulose, hemicelluloses, or lignin, which was indicated by the 2nd, 3rd, and 4th *Tp* as shown in Table 3 on low β. Results experiment of [18] strengthened this phenomenon that the reactions of IC with WP without moisture are diffusion-controlled, because the reactants of wood, including cellulose, hemicelluloses and lignin, are macromolecules. Moreover, some cellulose regions are highly crystalline, resulting in a lower mobility of these reactants.

IOP Conf. Series: Materials Science and Engineering **593** (2019) 012004 doi:10.1088/1757-899X/593/1/012004

Table 3. Data obtained from DSC thermograms including $T_o$, $T_p$, $T_e$ and $\Delta H$ at different β from measurement of sample made of mixture IC and WP at various ratios

| Sample | β (°C/mins) | Temperature (°C) | | | $\Delta H$ (J/g) |
|---|---|---|---|---|---|
| | | $T_o$ | $T_p$ | $T_e$ | |
| IC+5% WP | 2.5 | 225 | 279 | 291 | 80 |
| | | 288 | 304 | 311 | 47 |
| | | 311 | 324 | 350 | 102 |
| | | 350 | 409 | 446 | 168 |
| | 5 | 277 | 297 | 332 | 209 |
| | | 255 | 338 | 366 | 66 |
| | | 366 | 411 | 444 | 95 |
| | 10 | 292 | 327 | 384 | 189 |
| | | 361 | 421 | 464 | 97 |
| | 20 | 317 | 350 | 396 | 190 |
| | | 381 | 441 | 500 | 152 |
| IC+10% WP | 2.5 | 252 | 280 | 292 | 55 |
| | | 329 | 331 | 360 | 148 |
| | | 360 | 386 | 426 | 84 |
| | 5 | 294 | 305 | 325 | 121 |
| | | 278 | 325 | 365 | 67 |
| | | 365 | 411 | 447 | 103 |
| | 10 | 300 | 329 | 378 | 183 |
| | | 376 | 431 | 471 | 152 |
| | 20 | 314 | 351 | 397 | 190 |
| | | 374 | 436 | 493 | 137 |
| IC+12% WP | 2.5 | 252 | 280 | 292 | 57 |
| | | 329 | 331 | 364 | 156 |
| | | 364 | 390 | 430 | 86 |
| | 5 | 294 | 299 | 326 | 111 |
| | | 307 | 343 | 365 | 59 |
| | | 365 | 410 | 447 | 99 |
| | 10 | 284 | 326 | 376 | 188 |
| | | 340 | 416 | 472 | 124 |
| | 20 | 299 | 354 | 396 | 165 |
| | | 378 | 426 | 473 | 69 |
| IC+15% WP | 2.5 | 243 | 273 | 292 | 71 |
| | | 306 | 317 | 353 | 101 |
| | | 367 | 371 | 414 | 41 |
| | 5 | 289 | 302 | 316 | 90 |
| | | 316 | 330 | 364 | 43 |
| | | 376 | 393 | 425 | 21 |
| | 10 | 219 | 344 | 379 | 287 |
| | | 310 | 403 | 469 | 111 |
| | 20 | 285 | 355 | 394 | 216 |
| | | 357 | 419 | 466 | 69 |

For conclusions, *Tp* on the thermograms DSC indicated IC cured either using water or moisture WP. However when IC reacted with WP, the *Tp* emerged more than 2 which indicated IC also reacted with wood components. The reaction enthalpy and reaction rate were lower for WP than for the water indicated IC reacts with water in preference. Value of enthalpy seemed irregular and varies probably due to different structures and compositions of WP. This influence, however, was masked by the moisture present in the wood (free and bound water), because water is much easier to react with IC than were the –OH within the WP because of the low mobility of the hydroxyls in WP. In other words, IC may react with both –OH and water in IC/WP mixture. This led to the significant influence of the moisture content in WP on the curing kinetics, in terms of DSC thermograms and reaction enthalpy. It was suggested to calculate activation energy since the heating rates used up to 4 and to understand whether water, WP, or both affected the main reaction mechanism among IC, water within the wood (free and bound water), or wood components.

## 4. References

[1] Frazier C E 2003 *Handbook of Adhesive Technology 2nd Edition, Revised and Expanded* Chapter 33, ed A Pizzi and K L Mittal (New York: Taylor & Francis Group)
[2] www.grandviewresearch.com 2018
[3] Johns W E 1982 *J Adhes* **15** p 59
[4] Grøstad K and Pedersen A 2010 *J Adhes Sci Tech* **24** p 1357
[5] Kenji U and Etsuya Y 2013 Production method for particleboard and fibreboard *Eur Patent Appl* EP 2 666 609 A1
[6] Nuryawan A and Alamsyah E M 2018a *Applied Adhesive Bonding in Science and Technology* Chapter 5 (Croatia: Intechopen)
[7] Irle M 2018 Panels of the moon. *Wood Based Panels International* Feb/Mar 50
[8] Papadopoulos A N 2006 *BioResources* **1**(2) pp 201–208
[9] Dziurka D, Mirski R 2010 *Drvna Industrija* **61** (4) p 245
[10] Nuryawan A, Risnasari I, Sinaga P S 2009 *Jurnal Ilmu dan Teknologi Hasil Hutan* **2**(1) p 57
[11] Zhou XB, Frazier CE 2001 *Int J Adhes Adhes* **21** p 259
[12] Bao S, Daunch WA, Sun Y, Rinaldi PL, Marcinko JJ, Phanopoulos C 2003 *Forest Prod J* **53** p 63
[13] Frazier CE, Ni J 1998 *Int J Adhes Adhes* **18** p 81
[14] Rowell R M and Ellis W D 1981. *Bonding of isocyanates to wood. In Urethane Chemistry and Applications*, ed Edwards et al (Washington DC: ACS American Chemical Society Symposium Series) p 263–284.
[15] Weaver F W and Owen N L 1995 *Appl spect* **49**(2) p 171
[16] Umemura K, Takahashi A, Kawai S 1998 *J Wood Sci* **44** p 204
[17] Umemura K, Takahashi A, Kawai S 1999 *J Wood Sci* **74** p 1807
[18] He G, Yan N 2005 *Int J Adhes Adhes* **25** p 450
[19] Nuryawan A and Alamsyah E M 2018b *Thermal stability of isocyanate as particleboard's adhesive investigated by TGA (thermogravimetric analysis)*. (Medan: Paper presented in International Humanosphere Science Symposium) October 15
[20] BASF Corporation 2000 *Polyurethane MDI Handbook*
[21] Harper D P, Wolcott M P, Rials T G 2001 *Int J Adhes Adhes* **21** p 137
[22] Hatakeyama T and Quinn F X 1999 *Thermal Analysis Fundamentals and Applications to Polymer Science* (England: John Wiley & Sons)
[23] Mamiński M, Jaskŏlowski W, Parzuchowski P 2014 *Mediterr J Chem* **3**(3) p 877
[24] Wendler S L, Frazier C E 1996a *Int J Adhes Adhes* **16** p 179
[25] Wendler S L, Frazier C E 1996b *J Appl Polym Sci* **61** p 775

**Acknowledgments**
This study was funded by a Postdoctoral Research Scheme Grant of the year of 2017-2018 to AN from the Directorate for Research and Community Service, Ministry of Research, Technology and Higher Education, Republic of Indonesia. Further, AN thanks to Ms.Roslina Abdul Rashid for laboratory work at Institute Advance Materials, University Putra Malaysia.

IOP Conf. Series: Materials Science and Engineering **593** (2019) 012005 doi:10.1088/1757-899X/593/1/012005

# Structural and Mechanical Properties of Carbon-Carbon Biocomposites

**D Tahir[1*], S Sattar[1], S Fatimah[1], N Rauf[1], B Armynah[1]**

Department of Physics, Hasanuddin University, Makassar 90245, Indonesia

*Email: dtahir@fmipa.unhas.ac.id

**Abstract**. Carbon-carbon composites (C/C composites) have been synthesized from organic sawdust as a carbon filler material and carbon asphalt as matrix of composites. The structural and mechanical properties of carbon-carbon composites from sawdust-asphalt are focused in this experiment. The structural has been characterized by using FTIR, XRD and XRF. The mechanical properties has been resulted from hardness, strength, and attrition testing. According to XRF data, CaO is the main content from carbon filler in composite decrease with increasing carbon asphalt matrix and vice versa for $SO_3$ as the main content from carbon tar as a matrix. Changing structure of C/C composites indicated sawdust as carbon has been bound to asphalt as carbon composites. Beside this, adding carbon to carbon composites shows increasing hardness from 6,57 to 9,27 kg/mm$^2$, strength of sample from 11,90 to 18,56 N/mm$^2$, and decreasing attrition of material from 1,64 to 0,28 gr. According to the data shows that adding sawdust to asphalt makes increasing hardness and strength of C/C composites. It means that the C/C composites from organic material sources has a good material composites which has low degradation under high temperature, economically, and eco-friendly to apply in asphalt as material of road construction. It means that organic material very potential as a carbon-carbon composite materials for various applications.

## 1. Introduction

Carbon-carbon composites (C/C composites) are a family of composite materials which composed of fibrous substrates and chemically vapor deposited matrices [1]. The properties of this carbon are low density, high thermal conductivity and shock resistance, low thermal expansion and high modulus [2-4]. Carbon/carbon is mostly used in aerospace applications, mainly for aircraft disc brakes, and parts of rocket [1]. On the other hands, C/C uses as material coating to protective layers which will be impermeable to oxygen at 1000°C and above. Applying C/C composites to silicon carbide [5] and boron nitride [6] give incomplete oxidation protection. It has been shown that C/C composites are affective in protecting against oxidation in air at temperatures up to 1000°C. The large strength-to-weight ratio and the retention of strength at high temperature make C/C composites attractive at high temperature structural materials.

The unique properties of C/C composites have attracted attention from many researchers to develop and deeply study of carbon-carbon composites, including developing a material sources of C/C composites. Several researcher have used a different material sources of C/C composite because it tolerated by other materials [7], such as coal tar pitches [8] and carbon nanotubes in a polystyrene matrix [9].

The aim of this work is to produce C/C composites using organic as material sources that has higher hardness and strength, and low degradation. The characterization of C/C composites using by X-Ray Fluorencence (XRF), Fourier Transform Infra-Red (FTIR) and X-Ray Diffraction (XRD). The mechanical properties identified by hardness, strength, and attrition. In this experiment, we present a sustainable, inexpensive and eco-friendly approach of C/C composites.

## 2. Experimental

### *2.1. Preparation C/C composites*
Firstly, C/C composite makes from wood waste (C) as material source of carbon and tar (CC) as material source of carbon composite. In this experiment, there are 3 different composition of C/C composite. The ratio of C/C composite 80:20 where 9 gram of C and 3,5 gram of CC, 70:30 means 8,75 gram of C and 3,75 gram of CC, 60:40 where 8,5 gram of C and 4 gram of CC.

### *2.2. Characterization*
C/C composites have identified from structure and mechanical properties.

*2.2.1. Structure of C/C composites.* Characterized of physical properties by using X-Ray Fluorescence (XRF) to determining chemical composition, Fourier Transform Infra-Red (FTIR) to identifies chemical compounds and X-Ray Diffraction (XRD) to identification of a crystalline material and provide information on unit cell dimensions.

*2.2.2. Mechanical properties.* Mechanical properties of C/C composites contains three testing are:

- Hardness testing performed by using equipment based Brinell method. The result of this testing calculated by using formula below.

$$\mathrm{BHN}=\frac{2\mathrm{P}}{(\pi.\mathrm{D})(\mathrm{D}-\sqrt{D^2-d^2})} \quad (1)$$

- Strength testing was tested using tensile machine based pressure formula.

$$\mathrm{P}=\frac{\mathrm{F}}{A} \quad (2)$$

- Attrition testing conducted compositely on disc rotating. Attrition can be seen from the mass changed. Changing of mass obtained by weighing the mass of the C/C composite before and after testing.

## 3. Result and Discussion

### *3.1. Structure of C/C composites*
X-ray Fluorescence (XRF) spectrometry is a technique to determination of many major elemental compositions of earth materials. The composition of C/C composite determined by XRF shows on table below.

**Table 1.** Effect of carbon to carbon composite by X-ray Fluorescence spectrometry data.

| Oxide (wt. %) | Sample | | | | |
|---|---|---|---|---|---|
| | Wood Waste (C) | C:CC (80:20) | C:CC (70:30) | C:CC (60:40) | Tar (CC) |
| CaO | 58.16 | 29.13 | 26.37 | 23.12 | 0.00 |
| $K_2O$ | 24.08 | 23.75 | 22.42 | 20.25 | 0.00 |
| $SiO_2$ | 6.57 | 4.95 | 4.52 | 2.87 | 0.00 |
| $P_2O_5$ | 4.09 | 1.84 | 1.64 | 3.97 | 1.15 |
| $Fe_2O_3$ | 5.18 | 3.39 | 2.25 | 2.83 | 0.03 |
| $SO_3$ | 0.00 | 36.33 | 42.26 | 45.74 | 98.49 |
| LOI | 1.92 | 0.61 | 0.54 | 1.22 | 0.33 |

Wood waste as material source of carbon contains 58,16% of CaO and 24,08% of $K_2O$. Tar as material source of carbon composite contains 98,49% of $S0_3$. Adding carbon composites to carbon makes decreasing composition of CaO and $K_2O$ but increasing $SO_3$. It can be seen on different ratio of C/C composite respectively. Compositions of $SO_3$ on material C/C composites are 36,33%, 42,26%, and 45,74% on ratio 80:20, 70:30, and 60:40 respectively. Composition of CaO and $K_2O$ on material composites are 52,88%, 48,79%, and 43,37% with ratio 80:20, 70:30, and 60:40 respectively. This data shows that addition of carbon composites has impacted to the properties of carbon and vice versa. It indicated a material C/C composites can reduce degradation of tar properties. In the other side, composition of $SiO_2$ of carbon has decreased from 6,57% to 2,87% on ratio 60:40 of C/C composites.

Characterization of FTIR spectroscopy was used for identification of functional groups in material carbon from wood waste, C/C composites, and carbon composites from tar. As figure 1 shown, the curve carbon have a strong absorption peak at around 2924 $cm^{-1}$ [10] which is the absorption peak of O=H and nothing on curves of C/C composites. This is mainly due to $H_2O$ molecules involved in the process of synthesis carbon from waste wood. The peak in 2363 $cm^{-1}$ is attribute to C=C absorption peak of aromatic ring and C=O stretching vibration appear between 1456 $cm^{-1}$ [11] and 1561 $cm^{-1}$ [12] The peaks around at 1100 $cm^{-1}$ are attributed to aromatic C-H bending vibrations. However, the aromatic C-H absorption peak 1018 $cm^{-1}$ from carbon and at 1141 $cm^{-1}$ from C/C composites with ratio 60:40 are smaller than other curves. There may be having a polymerization reaction of compounds from carbon and carbon composites to form a multi-fused ring material [13]

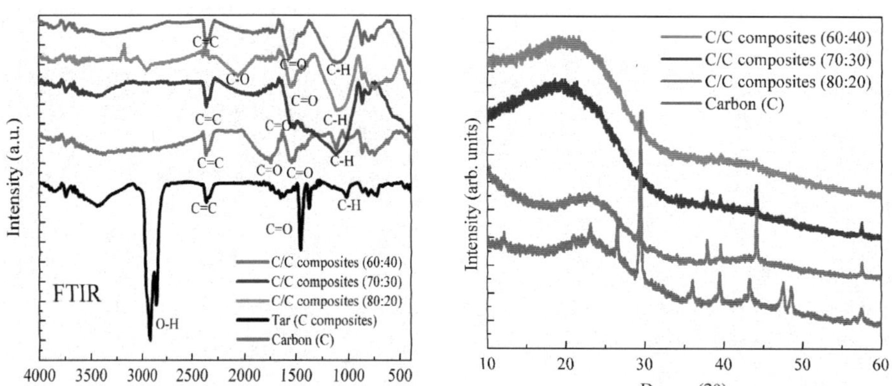


(a) (b)

**Figure 1.** Physical properties of C/C composites by (a) FTIR and (b) XRD data.

The effect of carbon composites for carbon that identified by FTIR on Figure 1a shows changing intensity at surface state area but not changed a moleculer bonding. FTIR data for C/C composites with ratio 80:20 shows instability condition at peak 2300 $cm^{-1}$ changing a moleculer bonding become C-O at

2062 $cm^{-1}$ due to the several carbonaceous particle not bonding with tar particle forming unstable bonding.

Figure 1b shows that carbon from wood waste has peak degree at 29,5 and 43,3. C/C composite with ratio 80:20 has peak degree at 29,2 and 44,0. C/C composite with ratio 70:30 has peak degree at 29,4 and 44,0. C/C composite with ratio has peak degree at 29,6 and 44,0. The intensity peak in XRD decreased with the increase of ratio carbon composite on carbon. Carbon from waste wood indicated a domination of amorphous structure than a crystallite. It impacts to structure of carbon as amorphous fully. It can be seen from peak of carbon reduce become amorphous structure for high amount of tar in C/C composite. This is not surprising due to the structure of tar is amorphous.

### *3.2. Mechanical properties of C/C composites*

The characterizations of mechanical properties consist are hardness, strength, and attrition. Purpose of this experiment to identify a physical and mechanical properties from material composite from organic sources.

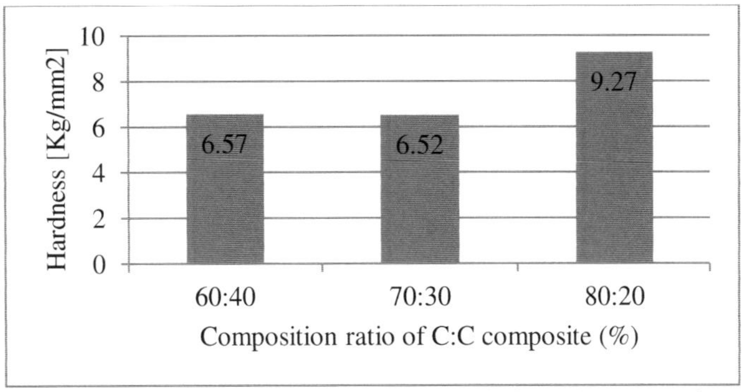


**Figure 2.** Graph of hardness testing result for C/C composites.

Changing composition of carbon-carbon composites impacts to hardness value. The result of hardness tenting of C/C composites shows on Figure 2. Composition ratio 60:40 has 6,57 $kg/mm^2$, 70:30 has 6,52 $kg/mm^2$ and 80:20 has 9,27 $kg/mm^2$. According to the diagram shows that increasing a composition of carbon comparable to hardness value. It means that a dominant materials from carbon wood waste more hardly than other less composition of carbon.

Decrease of hardness of mechanical properties on C/C composite caused by cracking core of material is not come out on the surface materials. A mechanical properties a materials composites was not depend on reinforcing and adhesive materials, but the characteristic of bonding between carbon and the composite that using on the material C/C composites. Interaction of their component impacts to mechanical properties for system carbon composites. The effect of surface treatment is significant on the mechanical properties of the resulting composites [1]

Result of strength testing shows that 11,90 $N/mm^2$, 12,02 $N/mm^2$, and 18,56 $N/mm^2$ with ratio composition are 60:40, 70:30, and 80:20 respectively. According to the graph below, it shows increasing composition of carbon comparable to the compressive strength of C/C composites.

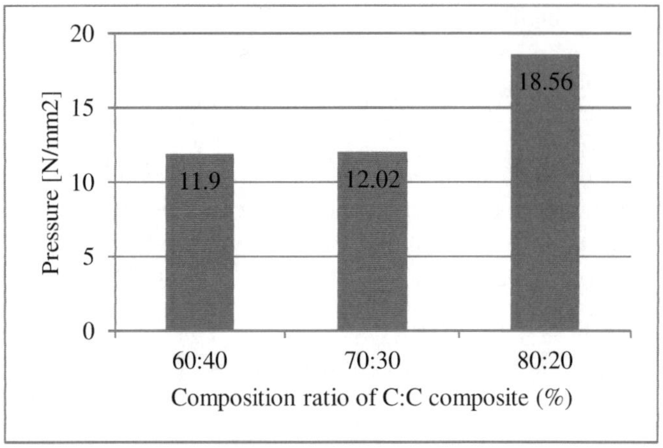


**Figure 3.** Graph of strength testing result for C/C composites.

The compressive strength not only depends to a micro-structure of material, such as pore and crack that occurs on carbonization process. The properties of carbon powder and deformations of materials on carbonization effects to the strength carbon composites. This caused likely other experiment that has reported [1, 14].

The attrition of sample was seen from mass changing before and after testing. Results of attrition testing with different ration of composition carbon-carbon composite at 60:40, 70:30, and 80:20 are 1.64 gram, 1.24 gram and 0.28 gram. It show on figure below.

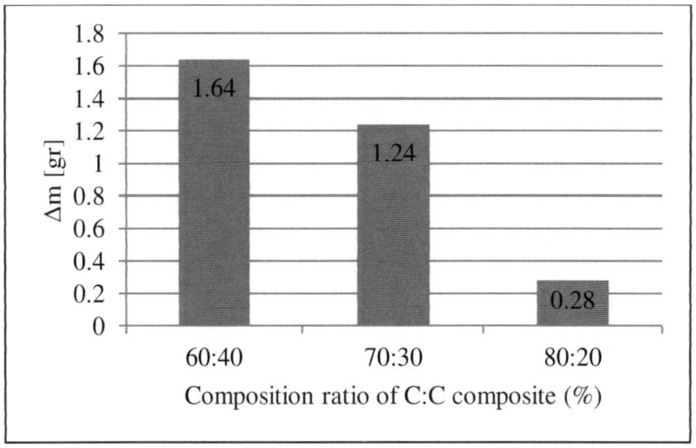


**Figure 4.** Graph of attrition testing result for C/C composites.

C/C composite with ratio 80:20 has low attrition at 0,28 gram. Changing micro structure of matrix carbon from densification process and thermal effect causes low attrition of material. Strong chemical bonds at the interface cannot provide a release mechanism, a weak interface adhesion resulted a crack. The processing cracks interphase initiated path from the interlaminar crack propagation [8].

## 4. Conclusions

Carbon-carbon composites based organic as material sources has been synthesized. According to analytical data from physical and mechanical properties shows that the level absorbance of molecule C/C composite increasing significantly. It has impacts to instability of moleculer bonding at C=C bonding and C-H bonding of C/C composites. Even thought detects a different of wavenumber from FTIR data, but still shows same bonding. On the other hands, mechanical properties of C/C composites shows that increasing composition of carbon comparable to increase hardness and strength value but decrease of attrition value. It indicated that organic material as material source of carbon is very potential to be C/C composites. This finding will lead to further research on applying C/C on industries.

**References**

[1] Windhorst T, Blount G 1997 *Mater. Des.* **18** pp 11–5
[2] McKee D W 1987 *Carbon* **25** pp 551 – 7
[3] Cheng L, Xu Y, Zhang L, Yin X 1999 *Carbon* **37** pp 977 – 81
[4] Smeacetto F, Salvo M, Ferraris M 2002 *Carbon* **40** pp 583–7
[5] Naslain R, Rossignol J, Hagenmuller O, Christin F, Heraud L, Choury J 1981 *Revue de chimie minerale* **18** p 544
[6] Hannache H, Quenisset J, Naslain R, Heraud L 1984 *J. Mat. Sci.* **19** p 202
[7] Fitzer E, Gadow R 1986 *Am. Ceram. Soc. Bull* **65** (2) p 326
[8] Chollon G, Siron O, Takahashi J, Yamauchi H, Maeda K, Kosaka K 2001 *Carbon* **39** (20) pp 65–75.
[9] Erik T Thostenson, Tsu W Chou 2002 *J. Phys. D: Appl. Phys.* **35** pp 77–80
[10] Gao X, Lu Y, Zhang R, He S, Ju J, Liu M, Lia L, Chen W 2015 *J. Mater. Chem. C* **3** pp 2302–9
[11] Sun D, Ban R, Peng H Zhang, Ge H Wu, Jian R Zhang, Jun J Zhu 2013 *Carbon* **64 (4)** pp 24–34.
[12] Zhu C, Zhai J, Dong S 2012 *Chem. Commun.* **48** p 9367.
[13] Dang A, Li H, Li T, Zhao T, Xiong C, Zhuang Q, Shang Y, Chen X, Ji X 2016 *J. Anal. Appl. Pyrol.* **119** pp 18-23
[14] Fitzer E, Huttner W 1981 *J. Phys. D: Appl. Phys.* **14** p 347

# Urban forest and financial resources perspective in Indonesia

**M Arsal[1*], D Arsal[2]**

[1] Muhammadiyah University of Makassar, Jl. Sultan Alauddin No 259 Makassar, Indonesia
[2] Research & Development Board South Sulawesi Jl. Urif Sumoharjo, No.269. Makassar

*E-mail: muryani@unismuh.ac.id

**Abstract** Urban forests can help to improve the environmental quality of urban areas because trees mitigating of urban heat island and adapting cities to environmental changes.The increasing urban population has a direct impact on urban natural capital, which is urbanization have been replaced green spaces with impervious materials. Urban forests in Indonesia is very limited and not well-managed due to the urban forest was made without any form of consensus with the people, but an initiative of the city government. Some of facing problems that are high land costs, diversity of land ownership, social and environmental pressures, and the spread of financing sources and other resources in urban forest cultivation and maintenance. In Indonesia, financial resources to maintain and management of urban forest it is not problem if users recognize and understand that benefit and function of urban forest. Local government, company, owners building, NGO's, local residents will participated in a partnership, but a key factor can influence of partnership is raise awareness and provide environment education; and legal aspect to supporting the partnership itself. This paper focuses on urban forests and financial resources in Indonesia. This paper concludes that urban forest management in Indonesia required a very large cost due to the high price of land, and they need participated of users and supporting legal aspect about who has responsibility of maintain and management of urban forest especially financial resources.

## 1. Introduction

Urban forests and trees help to improve the environmental quality of urban areas because trees mitigating of urban heat island and adapting cities to environmental changes, and provide cooling by evapotranspiration up to 7 KW [1]. Trees can absorb a variety of pollutants that are harmful to health. However, the increasing urban population has a direct impact on urban natural capital. One reason is that urbanization have been replaced green spaces with impervious materials. In summer, rising urban temperatures lead to increased electricity use, increased pollution, and increased discomfort among residents [2]. In Indonesia, urban forests is very limited and not well-managed due to the designation of urban forest was made without any form of consensus with the people, but rather an initiative of the city government [3]. Some of the problems that are often encountered are high land costs, diversity of land ownership, social and environmental pressures, and the spread of financing sources and other resources in urban forest cultivation and maintenance [4].

The main aim of this paper to find out of urban forest and financial resources to management urban forest in Indonesia. The structure of paper follows as described benefit and cost of urban forest and financial resources, methodology and conclusion.

### *1.1 Urban Forest*

Urban forests and all urban natural capital, such as urban agriculture, park and other urban greens can be considered as green infrastructure managed in one location [5]. Many studies have shown that forest-based benefits can be optimized through long-term management across the city to maximize the productivity of urban forests. Urban forests can only be fully planned and managed when urban residents and policy makers recognize and understand that all-round services that trees and green space provide. Environmental benefits of urban forest are often based on deferred costs; that is, without urban forests or trees, owners of property or governments will have to invest in additional engineering infrastructure or equipment to address environmental issues [5, 6]. For example, the canopy intercepts rainwater, which reduces the amount of water that falls to the ground and flows into the rainwater harvesting system, saving the city the cost of building larger capacity pipelines and storage facilities [7, 8]

Furthermore, urban forest can be benefit to human services, such as physical and mental conditions. Urban resident live a sedentary life, which increases the number of overweight or obese urban residents. The National Health in U.S conducts of individual activities levels and how to motivate people to participate in basic sports such as walking and cycling. Other studies have explored how to encourage urban design forms such as street layout, the presence of sidewalks and the proximity of parks activity. When the entire city or country gathers, the economic consequences of traditional light sports are enormous. The fatique patient after a short visit to the forest environment can enhance psychological and physical recovery, and access to the forest environment can be beneficial when suffering from fatigue [9]. Similarly, deferred costs are possible because people who are engaged in daily physical activity and exercise have lower medical costs.

All benefit of urban forest like environment benefit, human services requires the cost for planting, maintenance, materials and disposal. Potential energy savings in buildings by an urban tree planting program in California, they assuming total planting and stewardship costs of $ 2.5 billion or $ 50/tree [15], while cost of the maintenance trees is $1.9 million annual [6].

### *1.2 Financial Resources Perspective*

Economic assessment of urban forest services and functions into terms that increase public value. Urban forests are urban resource systems that can be cultivated and managed on all land in the city, including private and public property, as well as all socio-economic zones.

The return on investment is not easy to calculate. Industrial forests are used for market commodities to determine the price and income of resource products such as wood. In contrast, many products of urban forests are public goods. Multiple owners invest in the city's natural capital, producing products in the form of intangible features and benefits of each resident, visitor and user [10]. The trees provide $8.4 million in annual services and $1.9 million for maintenance, every $1 spent on tree management, residents receive $4.48 in benefits. Energy-saving value $6.20/tree, CO2 reduction $0.33/tree and atmospheric pollutant deposition $5.40/tree Large value associated with reduced stormwater runoff $47.80/tree and increased real estate value $144.70/tree [6].

Few private companies willing to invest in public goods, because the non-exclusive and unnatural conditions of the urban forest rarely generate profits. Government agencies have traditionally invested in public resources that are intuitively accepted by member of society as providing value, such as education or emergency response systems. Sustained political support of such investments is more likely if economic benefits can be demonstrated.

Users who live nearby may spend a small amount of money, while others may travel some distance, and their spending on meals, fuel, accommodation, and souvenirs can be proportionally based on the amount of time the park or forest visits [5].

Consumers claim that they are willing to pay about 9% to 12% for products in the downtown tree shopping district, rather than comparable areas without trees. Customer service, merchants' helpfulness and product quality are considered better by shoppers in the tree area [5]. While, trees provided a net annual benefit at $2.2 million in Modesto and $805,732 in Santa Monica [15]. The cost-benefit ratios for Modesto and Santa Monica are 1.85:1 and 1.52:1, respectively. For every $1 invested by the management, residents receive annual benefits of $1.85 and $1.52. Aesthetics and other benefits account for 50% to 80% of total annual revenue, while pruning expenses account for approximately 50% of total annual cost.

### *1.3 Urban Forest and Financial Perspective in Indonesia*

The developing countries including Indonesia, there is a critical challenge not only in terms of maintaining existing green areas, but urban forest management is also complicated due to lack of financial resources [11]. A few studies of urban forest related to financial and economic were conducted in Indonesia. Urban community is willing to pay for the maintain and cultivation of urban forests in Jakarta because it can bring many benefits, such as improved environment, play facilities, fishing places and strategic for sales places [12]. On the other hand, the local government cooperates with PT [13]. Pertamina to manage the urban forest Subang in Indonesia has been successful in growing crops of high economic value. Since 2012 land use of urban forest change was leisure activities, rice fields, residential areas and crop plantations. As one of the stakeholders, PT. Pertamina clearly prove the seriousness to do conservation program that have positive impacts on the quality of the environment. However, this conservation program faced a lot of obstacles and interference from outside parties who damage the vegetation.

This study showed the function of urban forests in Tangerang, Indonesia has not been fully benefited by the community because of the lack of trees that can be a shelter [14]. Only 5 % user of urban forest as strategic of sales place. In addition, this study suggests to to improving the function of urban forests such as the addition of vegetation to make urban forests more shady with lush trees, more intensive supervision and management in maintaining the condition of public green space so that they are always in good condition, and the awareness and responsibility of community in utilized urban forests by not damage or carrying out vandalism, maintain the cleanliness of the urban forest environment. There is no specific authority responsible for urban forest management due to the urban forest is not seen as an object to be managed intensively, thus some urban forests are almost without management [1]. This study found that only Hasanuddin University urban forest managed by itelft.

## 2. Methodology

The data were collected from secondary sources of journal publications and books and thus, analysed through understanding and synthesizing the literature.

## 3. Conclusion

In Indonesia, urban forests is very limited and not well-managed due to the designation of urban forest was made without participated of people, high land costs, diversity of land ownership, social and environmental pressures, and the spread of financing sources and other resources in urban forest cultivation and maintenance. Urban resident willing to pay for maintain and cultivation of urban forest in Jakarta. While, the local government cooperates with PT. Pertamina to manage the Subang urban forests in Tangerang has not been fully benefited due to the lack of trees and suggests to improving the function of urban forests (addition of vegetation, more intensive supervision and management in maintaining urban forest).

Financial resources to maintain and management of urban forest in Indonesia, it is not problem if users recognize and understand that benefit and function of urban forest such as environment benefit and human services, then they (local government, company, owners building, NGO's, local residents) will participated in a partnership. Lastly, a key factor can influence of partnership were raise awareness and provide environment education; and legal aspect to supporting the partnership itself.

**References**

[1] Riyadi D 2016 *Identify the socio-economic benefits of the Pesanggrahan Sangga Buana city forest for the surrounding community* – in Bahasa (Bogor: Bogor Agricultural University)

[2] Imansari N and Kadiyanta P 2015 Provision of urban forests and city parks as public green open spaces according to people's preferences in the downtown area of Tangerang – in Bahasa *Ruang* **1**(2) pp 101–10

[3] Arsal D and Johar F 2016 Challenges of urban forest development in Makassar city Indonesia *Imperial Journal Interdisciplinary Research* **2**(8) pp 1320–25

[4] Livesley S J, Baudinette B dan Glover D 2014 Rainfall interception and stem flow by eucalypt street trees – the impacts of canopy density and bark type *Urban Forestry & Urban Greening* **13**(1) pp 192–7

[5] Wolf KL 2004 Economics and public value of urban forests *Urban Agriculture Magazine, Special Issue on Urban and Periurban Forestry* **13** pp 31–3

[6] Song XP, Tan,PY, Edward P and Richard D 2018 The economics benefit and cost of trees in urban forest: A systematic review *Urban Forest & Urban Greening* **29** 162–170

[7] McPherson EG dan Simpson JR 2003. Potential energy savings in buildings by an urban tree planting programme in California. *Urban Forestry & Urban Greening* **2** (2) pp 73–86

[8] Song XP, Tan,PY, Edward P and Richard D 2018 The economics benefit and cost of trees in urban forest: A systematic review *Urban Forest & Urban Greening* **29** 162–170

[9] Rahman MA, Smith JG, Stringer P and Ennos AR 2011 Effect of rooting conditions on the growth and cooling ability of *pyrus calleryana Urban Forestry & Urban Greening* **10** (3) pp 185–192

[10] Hardin P J dan Jensen R R 2007 The effect of urban leaf area on summertime urban surface kinetic temperatures: A Terre Haute case study Urban *Forestry & Urban Greening* **6** (2) pp 63–72

[11] Barton P and Johnson M 2014 *Tree management and social housing in England, tree, people, and the building environment* II Conference Proceedings, Urban Trees Research Conference, University of Birmingham, Edgbaston UK pp 96–106

[12] Soares AL, Rego FC, McPherson EG, Simpson JR, Peper PJ, Xiao Q 2011 Benefits and costs of street trees in Lisbon Portugal *Urban Forestry & Urban Greening* **10** (2) pp 69–78

[13] Livesley S J, Baudinette B dan Glover D 2014 Rainfall interception and stem flow by eucalypt street trees – the impacts of canopy density and bark type *Urban Forestry & Urban Greening* **13**(1) pp 192–7

[14] Jones N, Collins K, Vaughan J, Benedikz T and Brosnan J 2005 *The role of partnership in urban forestry*. In urban forest and trees: A reference book. (Eds. Konijnendijk CC. Nilsson K, Randrup T, Schipperijn) pp 187–205 Springer

[15] Öström ES, Nordin M, Lundell Y, Dolling A, Wiklund U, Karlsson M, Carlberg B dan Järvholm LS 2014 Restorative effects of visits to urban and forest environments in patients with exhaustion disorder *Urban Forestry & Urban Greening* **13** (2) pp 344–354

# Phytochemical Analysis of Mangrove Leaves (*Rhizophora* sp.)

**Syahidah[1*], N Subekti[2]**

[1]Forestry Faculty, Hasanuddin University Jl. Perintis Kemerdekaan Km. 10 Tamalanrea, Makassar 90245, Indonesia
[2]Faculty of Mathematics and Natural Science, State University of Semarang, Central Java, Indonesia

*Email : syahidah@unhas.ac.id

**Abstract**. Mangrove plants have been reported as resources of many traditional folk of medicine and contain several kinds of trace elements which are being used to treat diseases. The aim of this study was to analyzed the phytochemical compounds of mangrove leaves. Our previous studies showed that the extract of mangrove leaves tend to be polar. The acetone and methanol extract of the mangrove leaves had high anti termite and antifungal activity. Based on this information, phytochemical content of mangrove leaves have been analyzed to find out the chemical substances that responsible for the bioactivity properties. Folin-Ciocalteu method was used to quantify the total phenolic content using a calibration curve of gallic acid, while for phytochemical were analyzed using spectrofotometric method. The results showed that the acetone extract of mangrove leaves containing alkaloid, polyphenolic, flavonoid, and total tannin higher than the methanol extract. This data will be further analyzed to obtain the compounds that may responsible for the biological performance of mangrove leaves.

## 1. Introduction

Mangroves are one of the most productive and unexplored ecosystem that approximately covers one fourth of the world coastline with high diversity of thriving organisms. Mangroves support the conservation of biological diversity for a number of endangered species by providing habitats, nurseries, nutrients, and spawning grounds [1]. Mangroves play also a key role in human sustainability and livelihoods, being heavily used for food, timber, fuel and medicine. They offer protection from catastrophic events, such as tsunami, tropical cyclones and tidal bores and can dampen shoreline erosion [2].

Mangroves contribute to numerous environmental services, including trapping and recycling organic matter, providing shelters and surfaces for terrestrial and aquatic organisms, and contributing to the overall health of coastal environments. The mangrove ecosystem represent an inter phase between terrestrial and marine communities, which receive a daily input of water from the ocean (tides) and freshwater, sediments, nutrients and silt deposits from upland rivers. Mangroves may grow as trees or shrubs according to the climate, salinity of water, topography and edaphic features of the area in which they exist. Mangroves are salt-tolerant evergreen forests found along sheltered coastlines, shallow-water lagoons, estuaries, rivers or deltas in 124 tropical and subtropical countries and areas [3][4].

Mangroves also are well known for their ecological importance and a rich source of several bioactive compounds such as steroids, triterpenes, saponins, flavonoids, alkaloids and tannins having therapeutic significance. These plants have been extensively used in traditional medicine and different studies have revealed their activity against human, animal and plant pathogens. The leaf and stem extracts of *S. maritima* are rich source of natural antioxidant with moderate antimicrobial activities. It has also been reported that these are an excellent source of antiviral compounds as compared to the seaweeds and sea grasses [5][6]. The ethno-pharmacological consequence pointed out that the mangrove plants are traditionally used for the treatment of rheumatism, painful arthritis, inflammation, asthma, antioxidant, free radical scavenging, anti- inflammatory, antinociceptive, diabetes and hepato-protective actions [7][8]. Apart from medicinal purposes, mangroves are also used as flavoring agents, in textiles, mats, paper, housing, baskets, boats and tapa cloth, tannins etc. [1].

Our previous studies showed that the extract of mangrove leaves tend to be polar. The acetone and methanol extract of the mangrove leaves showed high anti termite and antifungal activity [9].

## 2. Material and Methods

### *2.1. Material*

Mangrove (*Rhizophora* sp.) fresh and mature leaves was collected from the mangrove forest in Semarang, Central Java Province, Indonesia.

### *2.2. Extraction and Fractionation*

The mangrove leaves were dried and grounded in a hammer mill. The leaves mill (40-60 mesh) then extracted according to the procedure reported previously with slight modification [10]. In this study, leaves mill was used instead of small wood-pieces and the leaves mill was extracted with acetone first and the residue was then extracted again using methanol until the extract solution became colorless. These acetone and methanol extracts were then successively fractionated into n-hexane, ethyl acetate, and water to give their soluble fractions.

### *2.3. Phytochemical Screenings*

The acetone and methanol extracts were tested for the presence of alkaloids, steroids, tannins, saponins and terpenoids. The qualitative results are expressed as (+) for the presence and (-) for the absence of phytochemicals.

*2.3.1. Test for alkaloids.* Few mg (about 15 mg) of each extract was separately stirred with 1% HCl (6 mL) on a water bath for 5 min and filtered. These filtrates were divided into three equal parts.

a. Dragendorff's test: to one portion of the filtrate, Dragendorff's reagent (Potassium bismuth iodide solution) (1 mL) was added; an orange red precipitate shows the presence of alkaloids.
b. Mayer's test: to one portion of filtrate, Mayer's reagent (Potassium mercuric iodide solution) (1 mL) was added. Formation of cream colored precipitate gives an indication of the presence of alkaloids.
c. Wagner's test: Potassium iodide (2 g) and iodine (1.27 g) were dissolved in distilled water (5 mL) and the solution was diluted to 100 mL with distilled water. Few drops of this solution were added to the filtrate; a brown colored precipitate indicates the presence of alkaloids [11][12].

*2.3.2. Test for steroids and terpenoids.*

a. Salkowski test: The crude extract (about 100 mg) was separately shaken with chloroform (2 mL) followed by the addition of concentrated $H_2SO_4$ (2 mL) along the side of the test tube, a reddish brown coloration of the interface indicates the presence of terpenoid [13].
b. Liebermann-Burchard test: Each extract (100 mg) was shaken with chloroform in a test tube; few drops of acetic anhydride was added to the test tube and boiled in a water bath and rapidly cooled in iced water. Concentrated $H_2SO_4$ (2 mL) was added alongside of the test tube. Formation of a brown ring at the junction of two layers and turning the upper layer to green shows the presence of steroids while formation of deep red color indicates the presence of triterpenoids [11].

*2.3.3. Test for tannin.*

a. Preparation of a Standard Solution of 1000 ppm Tannic Acid
0.1 gram of tannic acid was dissolved in 100 ml of distilled water. This standard solution should be made new every will running a test [14]. The dilution series preparation were 4, 8, 12, 16, 20 ppm. Each 1 ml was taken from the dilution series and put in a 10 ml pumpkin container containing 35 ml of aquadestilata. In the flask, 0.5 ml of reagent folin type was added, left for 3 minutes and added 1.2 ml of saturated $Na_2CO_3$ 7.5% solution, incubated for 15 minutes. Then the absorption is read at a wavelength of 740 nm.

b. Determination of maximum absorption wavelength
Taken one of the concentrations of the standard solution, its absorption is measured in the wavelength range of 400-800 nm. The wavelength that shows the highest absorption value is the maximum wavelength. The standard curve is made by connecting the concentration of standard solutions with the absorption results obtained from measurements using a UV-VIS spectrophotometer at a wavelength of 740 nm.

c. Determination of Tannin Levels
0.5 gram of maserate is weighed and dissolved with aquadest until reach 10 ml volume. If the sample hasn't dissolved completely, it can be apply a device to homogenize the solution. Carefully piped 1.0 ml of sample, put in a 10 ml container containing 7.5 ml of aquadest. 0.5 ml of Folin Denis reagent was added, left to stand for 3 minutes, added 1.0 ml of saturated Na2CO3 solution. Incubated for 15 minutes, then read up at maximum wavelength. Calculated using the standard curve that has been obtained so that the concentration of the measured sample is known.

*2.3.4. Test for saponin.* Each of extracts (0.5 g) was separately shaken with distilled water (10 mL) in a test tube. The formation of frothing, which persists on warming in a water bath for 5 min, shows the presence of saponins [15].

*2.4. Determination of total phenolic contents (TPC) and total flavonoid contents (TFC)*

*2.4.1. Total Phenolic Content.* Total phenolic content was analyzed using the Folin–Ciocalteu colorimetric method with some modifications[16][17]. An aliquot of 0.3 mL of extract was mixed with Folin–Ciocalteu phenol reagent (2.25 mL). After 5 min, 6% sodium carbonate (2.25 mL) was added and the mixture was allowed to stand at room temperature for 90 min. The absorbance of the mixture was measured at 725 nm. Standard calibration curve for gallic acid in the range of 0–200 lg/mL was prepared in the same manner and results were expressed as mg gallic acid equivalent (GAE) per gram of extract.

*2.4.2. Total Flavonoid Content.* Total flavonoid content was determined using the aluminum colorimetric method [18][19] with some modifications using quercetin as the standard. A calibration curve of quercetin was prepared in the range of 0–200 lg/ mL. Briefly, extract (0.5 mL) and standard (0.5 mL) were placed in different test tubes and to each 10% aluminum chloride (0.1 mL), 1 M potassium acetate (0.1 mL), 80% methanol (1.5 mL) and distilled water (2.8 mL) were added and mixed. A blank was prepared in the same manner where 0.5 mL of distilled water was used instead of the sample or standard, and the amount of aluminum chloride was also replaced by distilled water. All tubes were incubated at room temperature for 30 min. The absorbance was taken at 415 nm. The concentration of flavonoid was expressed as mg quercetin equivalent (QE) per gram of extract.

## 3. Results and Discussion

*3.1. Phytochemical Screening*

The results of the phytochemical test carried out on the extracts were recorded as shown in Table 1. Preliminary phytochemical screening revealed the presence of alkaloid, flavonoids, phenolic, and tannin in both of the acetone and methanol extracts. On the contrary, terpenoid, steroid and saponin were absent

IOP Conf. Series: Materials Science and Engineering **593** (2019) 012007 doi:10.1088/1757-899X/593/1/012007

in both of the extracts. Alkaloid were much abundant in acetone crude extract according to Dragendorf and Wagner methods, but less abundant according to Mayer method. Alkaloid was found less abundant in methanol crude extract according to Wagner method, but found minute according to Dragendorf and Wagner methods.

Phytochemical constituents in the various part of the plant vary significantly. Several medicinal plants are used in traditional medicines for curing many diseases. These plants have been extensively used in traditional medicine and different studies have revealed their activity against human, animal and plant pathogens. The leaf and stem extracts of *S. maritima* are rich source of natural antioxidant with moderate antimicrobial activities. It has also been reported that these are an excellent source of antiviral compounds as compared to the seaweeds and sea grasses [5][6]. The study revealed that flavonoids and phenol were found to be present the extract. Several studies reported that, flavonoids show anti-allergic, anti-inflammatory, anti-microbial and anti-cancer activity. Tannin and phenol are necessary for the animal body for repair and maintenance. Flavonoids are now considered as an indispensable component in a variety of nutraceutical, pharmaceutical, medicinal and cosmetic applications. This is attributed to their anti-oxidative, anti-inflammatory, anti-mutagenic and anti-carcinogenic properties coupled with their capacity to modulate key cellular enzyme function [20].

**Table 1.** Phytochemical screening of acetone and methanol extracts of mangrove leaves

| **Extract** | **Alkaloid** | | | **Terpenoid and Steroid** | **Saponin** |
|---|---|---|---|---|---|
| | Mayer | Dragendorf | Wagner | | |
| Acetone extract | ++ | +++ | +++ | - | - |
| Methanol Extract | + | + | ++ | - | - |

*Legend: +++ (Much abundant); ++ (Less abundant); + (Minute); - (Absent)*

The phytochemical compounds detected are known to have medicinal properties. For example, alkaloids have been reported as powerful poison and many alkaloids derived from medicinal plants show biological activities like, anti-inflammatory [21], antimalarial [22], antimicrobial [23], cytotoxicity, antispasmodic and pharmacological effects [24][25]. Similarly, steroids derived from plants are known to have cardiotonic effect and also possess antibacterial and insecticidal properties [26]. They are very often used in medicines due to their well-known biological activities. Tannins, according to research, are known to have antibacterial [27], antitumor and antiviral activities ([28]. These alkaloids show bioactivity against Gram-positive bacteria and cytotoxicity against leukemia and HeLa cell lines [29]. Alkaloids, flavonoids and xanthones that are potent inhibitors of various oxidative processes in both in vitro and in vivo system. These phytochemical compounds identified in the extracts may be responsible for the biological activities of the mangrove leave extract.

*3.2. Total Phenolic Contents (TPC) and Total Flavonoid Contents (TFC)*

The results as shown in Table 2 revealed that the acetone extract of mangrove leaf has higher values in phenolic, flavonoid and tannin contents compared to the methanolic extract. The total phenolic content was different when extracted with different solvents [30]. Plant polyphenols are the significant group of compounds acting as free radical scavenging or primary antioxidants; therefore, it is justifiable to determine phenolic content in plant extract. Polyphenolic compounds have an aromatic benzene ring with substituted hydroxyl groups, including their functional derivatives. These are able to absorb free radicals and can chelate metal ions that could catalyze formation of ROS which promotes lipid peroxidation. Among polyphenols, flavonoids are of great importance because they help human body to fight against diseases. The ability of flavonoids to act as potent antioxidants depends on their molecular structures, the position of the hydroxyl group and other features in its chemical structure. They are abundantly found in plants as their glycoside [31]. The most abundant flavonol which has a good antioxidant property is quercetin, as it has all the right structural features for free radical scavenging activity [32].

Table 2. Total Phenolic Contents (TPC), Total Flavonoid Contents (TFC) and Total Tannin Contents (TTC) of acetone and methanol extracts of mangrove leaves

| Extract | TPC (mg GAE/g) | TFC (mg QE/g) | TTC (%) |
|---|---|---|---|
| Acetone extract | 10.024 | 2.641 | 6.703 |
| Methanol extract | 2.225 | 1.998 | 2.736 |

Tannin and polyphenols extracted from mangal species have also been reported to have cytotoxic, anti-neoplastic, antibacterial and anti-helmintic activities. Saponin and limonoids are reported to possess antimicrobial, inflammatory, cytotoxic, antifeedant and growth regulatory activity against a number of pathogens [1].

Mangroves have been a source of several bioactive compounds. Secondary metabolites like alkaloids, phenolics, saponins, flavonoids, steroids and terpenoids have been characterized from mangroves and have been tested successfully for toxicological, pharmaceutical and ecological importance. Studies on the bioactive compounds of mangrove plants often lead to the discovery of new therapeutic agents. The mangrove plants possess a number of biological activities such as antibacterial, antioxidant, anticancer, cytotoxic, antiproliferative, insecticidal, antimalarial, antifungal, antifeedant, antidiarrheal, central nervous system depressant, antimitotic, antileukemic and anti-plasmodial activities [33][34].

**References**

[1] Gouda S, G. Das, J K Patra 2015 *Mangroves: A Rich Source of Natural Bioactive Compounds. Recent Advances in Natural Products* (USA: Studium Press LLC)

[2] Carugati L, B Gatto, E Rastelli, M L Martire, C Coral, S Greco, R Danovaro 2018 *Impact of mangrove forests degradation on biodiversity and ecosystem functioning*. Scientific Reports (2018) 8:13298. www.nature.com/scientificreports/

[3] Hakim, L, D Siswanto, N Nakagoshi 2017 Mangrove Conservation in East Java: The Ecotourism Development Perspectives *The J of Tropical Life Sci.* **7**(3) pp 277–285

[4] Blankespoor B, S Dasgupta, G M Lange 2016 Mangrove as Protection from Storm Surges in a Changing Climate Policy Research Working Paper 7596 World Bank Group Development Data Group Surveys and Methods Team & Development Research Group Environment and Energy Team, March 2016.

[5] Patra J K, N K Dhal, H N Thatoi 2011 In vitro biaoctivity and phytochemical screening of Suaeda maritima (Dumort): A mangrove associate from Bhitarkanika, India *Asian Pacific Journal of Tropical Medicine* (2011) 727–734

[6] Sahoo, G, N S S Mulla, Z A Ansari, C Mohandass 2012 Antibacterial Activity of Mangrove Leaf Extracts against Human Pathogens. *Indian J. of Pharm. Sci.* **74** (4) pp 348–351

[7] Roome T, Dar A, and Naqvi A 2011 Evaluation of antinociceptive effect of Aegiceras corniculatum stems extracts and its possible mechanism of action in rodents *Journal of Ethnopharmacology* **135** pp 351–358

[8] Prabhakaran J and D Kavitha 2012 Ethnomedicinal importance of mangrove species of Pitchavaram. *International Journal of Research in Pharmaceutical and Biomedical Sciences*, **3**(2) pp 611–614

[9] Syahidah and N Subekti 2018 *Biological Activity of Mangrove Leaves Extract (Rhizophora sp.)*. (Makassar, Indonesia: Proceeding of The 1st Biennial Conference of Tropical)

[10] Wang S Y, P F Chen, S T Chang 2005 Antifungal activities of essential oils and their constituents from indigenous cinnamon (*Cinnamomum osmophloeum*) leaves against wood decay fungi. *Bioresour. Technol.* **96** pp 813–818

[11] Joshi A, Bhobe M, Saatarkar 2013 A Phytochemical investigation of the roots of Grewia microcos Linn. *J. Chem. Pharm. Res.* **5** pp 80– 87

[12] Abdullahi M N, Ilyas N, Ibrahim H 2013 Evaluation of phyto- chemical screening and analgesic activity of aqueous extract of the leaves of Microtrichia perotitii dc (Asteraceae) in mice using hotplate method. *Med. Plant Res.* **3** pp 37–43

[13] Ayoola G A, Coker H B, Adesegun S A, Adepoju-Bello A A, Obaweya K, Ezennia E C, Atangbayila T O 2008 Phytochemical screening and antioxidant activities of some selected medicinal plants used for malaria therapy in southwestern Nigeria. *Trop. J. Pharm. Res.* **7** pp 1019–1024

[14] Cunnif P 1996 *Official Method Of Analysis Of AOAC International sixteeth edition Vol II*, (USA: AOAC international Suite 500, 481 North freederick Avenue Gaithersburg: Maryland 20877-2417)

[15] Banso A, Adeyemo S, 2006 Phytochemical screening and anti- malarial assessment of Abutilon mauritianum, Bacopa monnifera and Datura stramonium. *Biokemistri* **18** pp 39–44

[16] Cai Y Z, Luo Q, Sun M, Corke H, 2004 Antioxidant activity and phenolic compounds of 112 traditional Chinese medicinal plants associated with anticancer. *Life Sci.* **74** pp 2157–2184

[17] Chlopicka J, Pasko P, Gorinstein S, Jedryas A, Zagrodzki P 2012 Total phenolic and total flavonoid content, antioxidant activity and sensory evaluation of pseudocereal breads. LWT – *Food Sci. Tech.* **46** pp 548–555

[18] Chang C, Yang M, Wen H, Chern J 2002 Estimation of total flavonoid content in propolis by two complementary colorimetric methods *J. Food Drug Anal.* **10** pp 178–182

[19] Stankovic M S 2011 Total phenolic content, flavonoid concentra- tion and antioxidant activity of *Marrubium peregrinum* L. extracts. *Kragujevac J. Sci.* **33** pp 63–72

[20] A N Pache, A D Diwan, S R Chandra 2016 Flavonoids: An overview. *J Nutr Sci* **2016** (5) pp 1–15

[21] Augusto L S, Josean F T, Marcelo S, Margareth F M, Petronio F A, Jose M B 2011 Anti-inflammatory activity of alkaloids: an update from 2000 to 2010 *Molecules* **16** pp 8515–8534

[22] Dua V K, Gaurav V, Bikram S, Aswathy R, Upma B, Dau D A, Gupta N C, Sandeep K, Ayushi R 2013 Anti-malarial property of steroidal alkaloid conessine isolated from the bark of Holarrhena antidysenterica. *Malaria J.* **12** pp 1–6

[23] Benbott, A., Yahyia, A., Belaı¨di, A., 2012. Assessment of the antibacterial activity of crude alkaloids extracted from seeds and roots of the plant Peganum harmala L. J. Nat. Prod. Plant Resour. 2, 568–573.

[24] Ameyaw, Y., Duker-Eshun, G., 2009. The alkaloid contents of the ethno-plant organs of three antimalarial medicinal plant species in the eastern region of Ghana. Int. J. Chem. Sci. 7, 48–58.

[25] Thite S V, Chavan Y R, Aparadh V T, Kore B A 2013 Preliminary phytochemical screening of some medicinal plants. *Int. J. Pharm. Chem. Biol. Sci.* **3** pp 87–90

[26] Alexei Y B, Joseph I S, Olga V F 2009 Endogenous cardiotonic steroids: physiology, pharmacology and novel therapeutic targets. *Pharmacol. Rev.* **61** pp 9–38

[27] Hisanori A, Kazuyasu F, Osamu Y, Takashi O, Keiji I 2001 Antibacterial action of several tannins against Staphylococcus aureus. *J. Antimicrob. Chemother* **48** pp 487–491

[28] Kumari M, Jain S 2012 Review paper, Tannins: an antinutrient with positive effect to manage diabetes. Res. *J. Recent Sci.* **1** pp 70–73

[29] Omar S, Chee C L, Fasihuddin A, Ni J X, Jaber H, Huang J, Nakatsu T 1992 Phenanthrene lactams from Goniothalamus velutinus. *Phytochem* **31** 4395–4397

[30] Abdelwahab S I, Abdul A B, Elhassan M M, Mohan S, Al- Zubairi A S, Mariod A A 2009 Antimicrobial and free radical scavenging activities of dichloromethane extract of Goniothalamus umbrosus. *Int. J. Trop. Med.* **4** 32–36

[31] Rajanandh M G, Kavitha J 2010 Quantitative estimation of b- sitosterol, total phenolic and flavonoid compounds in the leaves of Moringa oleifera. *Int. J. Pharm Tech Res.* **2** pp 1409–1414

[32] Kalita P, Barman T K, Pal T K, Kalita R, 2013. Estimation of total flavonoids content (TFC) and antioxidant activities of methanolic whole plant extract of Biophytum sensitivum linn. *J. Drug Delivery Ther.* **3** pp 33–37

[33] Patra J K and Thatoi H N 2011 Metabolic diversity and bioactivity screening of mangrove plants: A review. *Acta Physiologiae Plantarum* **33** pp 1051–1061

[34] Thatoi H N, Patra J K and Das S K 2011 Free radical scavenging and antioxidant potential of

mangrove plants: a review. *Acta Physiologiae Plantarum* **36**(3) pp 561– 579

IOP Conf. Series: Materials Science and Engineering **593** (2019) 012008 doi:10.1088/1757-899X/593/1/012008

# Supply Chain Management of the Plywood Industry in Indonesia

**Makkarennu[1]*, H R Natsir[1], Supratman[1]**
[1]Faculty of Forestry, Hasanuddin University, Jalan Perintis Kemerdekaan Km. 10, Makassar 90245

*Email: makkarennu@unhas.ac.id

**Abstract.** The plywood industry in Indonesia experienced a heyday in the 1980s and experienced ups and downs in the following year. One of the causes of the decline in plywood production in several industries is the supply chain management of raw materials. This study aims to determine supply chain mechanisms related to the flow of product, flow of financial and flow of information at the plywood industry in South Sulawesi. Data collection techniques are carried out with initial observations, field surveys, interviews, and literature studies. Data analysis was performed using qualitative and quantitative descriptive analysis. The results show that the supply chain in plywood mills has a fairly complex link starting from product flow, financial flow and information flow related to raw material supply, inventory and sales. This company needs to implement an integrated supply chain management to help maintain the sustainability of plywood production so as to increase the company's performance.

## 1. Introduction

Industry is an economic activity processing of raw materials, semi-finished goods into goods that will have a higher added value. Based on the classification of its productivity, the industry can be classified into producing goods and services. One of the producer industries is plywood industry [1]. Plywood has been the popular of Indonesia's processed wood industry products for several years. The highest export figures ever achieved in 1992 accounting to 9.7 million $m^3$ [2]. Plywood is the superior product in increasing national revenues, caused by production and exports experienced a rapid increase. Before the crisis period, plywood and other processed wood was the prime of non-oil exports, caused by the export value has increased every year [3].

The plywood industry experienced rapid growth during the period 1980-1997. This was related to the issuance of the Joint Decree of the Three Ministers (Agriculture, Ministry of Agriculture, and Trade) related to the ban on exporting of logs which is done gradually. The government effort to create added value of export wood product through to ban of log export is the government's anticipatory step in preventing there was a decline in foreign exchange due to the decline in world oil prices at the time. This has caused the coating industry developed a rapid growth [4]. However, the rate of Plywood production showed a downtrend with fluctuation. The number of plywood production in 2014, 2015, 2016 was at 3.579,113 $m^3$; 3,640,631 $m^3$ and 3,636,058 $m^3$ respectively [5].

One of the reasons for declining plywood production was the lack of logs raw materials [6]. Based on the above, this can result in a lack of availability in terms of raw materials. Inventory is one of the important stimulus supply chain due to changes in inventory policy can drastically change the level of responsiveness and supply chain efficiency [7]. The reduced production capacity of this plywood industry is one of them caused by inefficient supply chain management arrangements. Efficient supply chain management can be achieved if the management and supervision of distribution channel relationships was carried out cooperatively by all parties involved. The form of supply chain regulation really needs special attention [8].

Nowdays, supply chain management is no longer considered a new thing for the company. Many companies have implemented supply chain management in their business to increase the efficiency of the company's logistics process [9]. Supply chain is a concept where there are regulatory systems relating to product flow, information flow and financial information [10]. This flow is very important to identify related to the many aspects or elements in the supply chain in order to maintain product quality and availability. This study focuses on the flows in the supply chain including product flow, information flow, and financial flows in one of the plywood companies in South Sulawesi, Indonesia .

## 2. Research Methods

### *2.1. Study Site*

PT. KTC is located in Makassar City, it is one of the largest plywood companies in South Sulawesi. The consideration of choosing Makassar as the location of the plywood factory because of the availability of cheap labor, the port transportation is quite good and easy, and the transportation lines of the factory are quite cheap and smooth. In line with the times and technology in terms of production, PT. KTC has conducted a comprehensive restructuring in 2004, both related to production machinery and human resources by conducting employee training programs. To deal with market dynamics, PT. KTC strives to maintain a *brand image* that is already well known in the international market as well as through the improvement of product and service quality in *punctuality*. Improving the quality of production and service is supported by the improvement of skills and work capabilities of employees through continuous education and training both internally and externally. At present the number of employees working at PT. Katingan Timber Celebes is 1,578 people, consisting of 729 men and 849 women.

### *2.2. Data collection methodology and data analysis*

Data collection techniques were carried out through initial observation, field surveys, interviews and literature studies. The initial observation was the activity carried out with introduction of the location of the research which is to find out the state of PT. KTC based on information from various sources, especially from research results. Field surveys are conducted by visiting companies ranging from unloading, log pond, log yard, processing of raw materials and products, packaging, warehouses, and administration section. In addition, in-depth interviews conducted with informants in each unit of work are the production planning inventory control (PPIC) responsible for managing the inventory of raw materials, the purchasing that is in charge of the procurement of raw material preparation for the company's needs, production sector that is in charge of carrying out production activities and

IOP Conf. Series: Materials Science and Engineering **593** (2019) 012008 doi:10.1088/1757-899X/593/1/012008

supervising the quality of products and marketing section is responsible for generating revenue for the company by selling the company's products. Literature studies are conducted with research literature reviewers, journals, and reports related to research.

Data analysis was carried out by qualitative and quantitative descriptive analysis. Qualitative descriptive analysis is carried out to describe the relationship between supply chains starting from procurement, raw materials, production processes, distribution and marketing of plywood products, especially those related to the flow of products, information and finance. Quantitative descriptive analysis was carried out to calculate round wood supply starting from the initial inventory, obtaining raw materials, using raw materials and ending supplies of logs.

## 3. RESULTS

*3.1 Supply Chain in Plywood Industry PT. KTC*

Flow patterns in the plywood supply chain at PT. KTC (Figure 1) shows that there are three streams in the pattern divided into product flow, financial flow and information flow. The product flow begins from upstream to downstream, from the source of logs raw materials to consumers. The financial flow flows from downstream to upstream, that is, from consumers of plywood to source of logs while Information flow flows in the chain reciprocally.

IOP Conf. Series: Materials Science and Engineering **593** (2019) 012008 doi:10.1088/1757-899X/593/1/012008

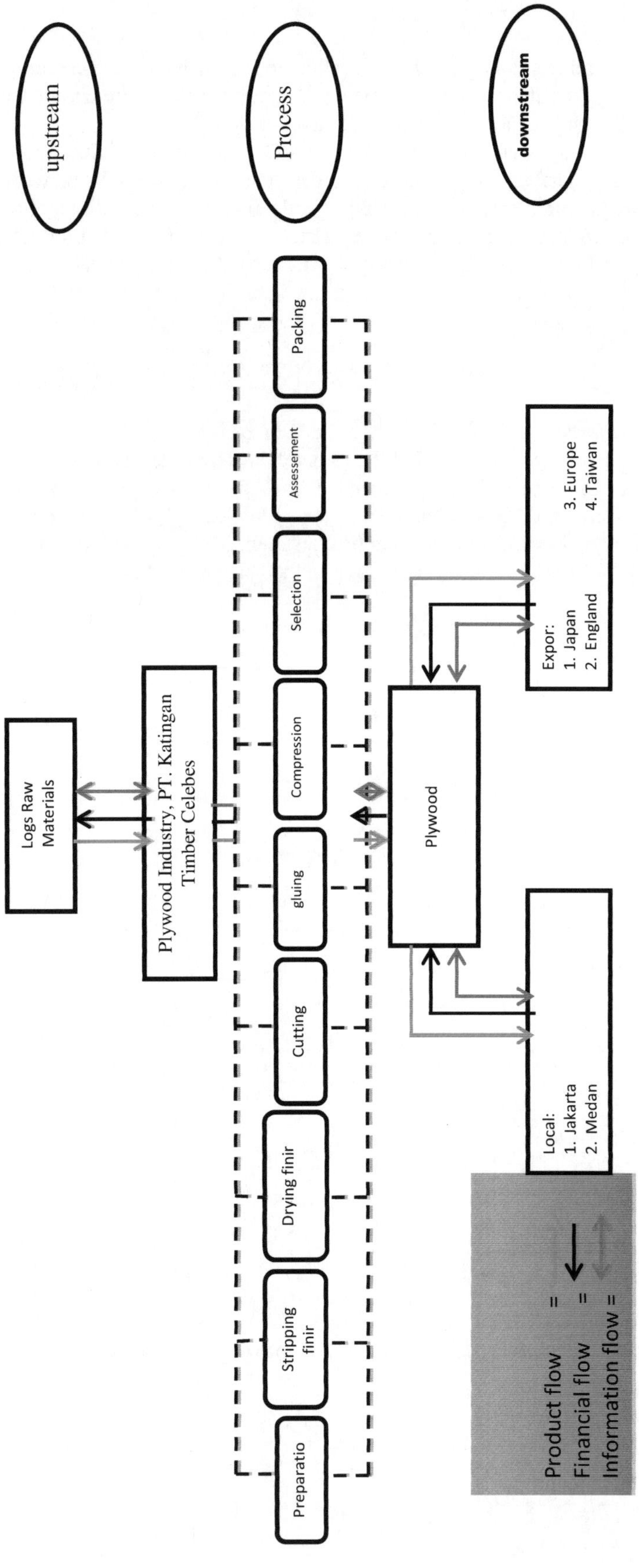


**Figure 1.** Plywood Supply Chain at PT. Katingan Timber Celebes

*3.1.1. Flow of Product*

The product flow starts from the sources of logs raw material, the production process of plywood until the plywood products are marketed to consumers. Logs raw material obtained originating from forest concession (own concession and other concession). The raw material obtained from own concession comes from PT. Gema Hutani Lestari , PT. Gorontalo Citra Lestari , and PT. Gorontalo Nusantara Jaya . Whereas other concession comes from PT. Nusaina Agro Tanah Merah Manise, PT. Dewata Wana Lestari, and PT. Wijaya Kencana .

Logs raw material which is from forest concession are placed in the Tallo River Estuary, which is an anchorage and unloading point for 4-5 days and then withdrawn to a log yard. The unloading point of the Tallo River estuary is only a place for unloading logs, not a place to collect or store logs, there is any facility to the port. Every day when unloading logs is immediately withdrawn to the log pond on the Tallo River. Logs that have been dismantled into the sea will be pulled to log pond 1 using a tug boat. Next from log pond 1, the wood is taken to log ponds 2, 3, and 4. Log pond 4 is the last log pond and is located at the area of the PT. KTC.

The production of plywood carried out through several sections at the start of the log handling, veneer reeling, veneer drying, veneer clipper, veneer jointing, glue spreading, pressing, grading, and packing. Furthermore plywood products then distributed to the main market, it is export markets.

*3.1.2. Flow of Finance*

Financial flow is a flow from downstream to upstream. Financial flows, begins from consumers to sources of raw materials. The financial flow referred to relates the system of purchasing, delivery, and selling the plywood. Buyers from abroad who want to buy plywood product purchasing with delivery order system. Delivery order is an activity to send product with the stages of starting the order of goods until the goods arrive at the customer with a Standard Operating Procedure that has been made and determined by a company with the policies of each company. As for product shipping costs, PT. KTC uses the Cost and Freight system where product transportation costs are borne by company. The company reaches the nearest port to the buyer. This fee includes insurance, taxes and other fees for removing goods from the port ware house. The selling prices of plywood products vary depending on the countries are of interest. The price of plywood products is exported to Japan and Taiwan, is \$ 600/m $^3$, while the European countries and the United Kingdom respectively are \$ 750 /m $^3$ and \$ 900 /m $^3$.

*3.1.3. Flow of Information*

Information flow is a flow that occurs both from upstream to downstream and from downstream to upstream. Information flow in the supply chain is related to the system of procurement of raw materials, the amount of raw material inventory, the amount of logs used, marketing, as well as the number of products exported and which are marketed locally.The process of procuring logs is handled by the log supply section. Procurement of logs is carried out using a contract system and is carried out after knowing a number of raw material requirements for production. The purchasing system uses a one price system, which is party purchases where the logs for one party are the same price and no longer pay attention to the type of logs purchased. If the inventory level dropped to the level of 9309.42 m$^3$ then the company needed to reorder [11]. Log orders made with the Reorder Point Approach are anticipated to avoid shortages and excess logs.

Furthermore, the company also actively explores cooperation with a number of forest concession and other suppliers who are considered capable of providing raw materials in order to ensure the continuity of production and avoid the occurrence of shortages of raw material inventories. In the process, the company will send a grader to check the quality and quantity of logs owned by the forest concession. If the logs inspected are deemed feasible, then the agreement of the contract agreement is immediately carried out by both parties, which in turn the supplier supplies logs for a certain period of time.

The fulfillment of logs at PT. KTC comes from natural wood and plantation wood. Figure 2 showed that the fulfillment of raw materials from natural wood decreased dramatically, accounted 605,623, 5 $m^3$ in 2014 to 47,120, 56 $m^3$ in 2016. However, in 2017 the fulfillment of raw materials from natural wood increased by 38,629, 63 $m^3$ from the previous year or by 45%. Fulfillment of raw materials from wood experienced a very significant increase from 2014 and 2015 which was at 33, 67 $m^3$ and 668, 96 $m^3$ respectively. However, in 2016 until 2017 the raw materials of plantation wood continued to decline from 302, 25 $m^3$ to 134, 37 $m^3$.

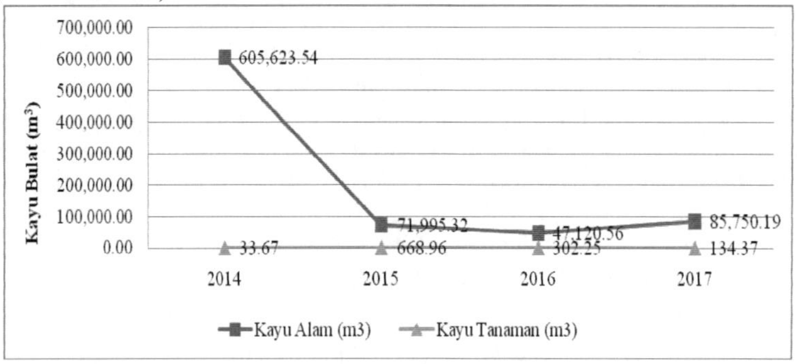


**Figure 2.** Fulfillment of Raw Materials for 2014-2017 ($m^3$)

Information flow discusses the number of round logs. Figure 3 shows that the use of logs in 2014 amounted to 100,526.89 m3 and continued to decline until 2016 until the use of logs was 47,785.81 m3. However, the use of round wood increased by 43.32% from the previous year.

The use of logs from plantation forest has increased significantly from 40.53 $m^3$ in 2014 to that of 668.96 $m^3$ in 2015. However, in 2016 to 2017 the use of logs fell down from 328.05 $m^3$ to 134.37 $m^3$. The comparison of the use of logs from the raw materials of natural wood and plants in 2014-2017 ($m^3$) can be seen in Figure 3.

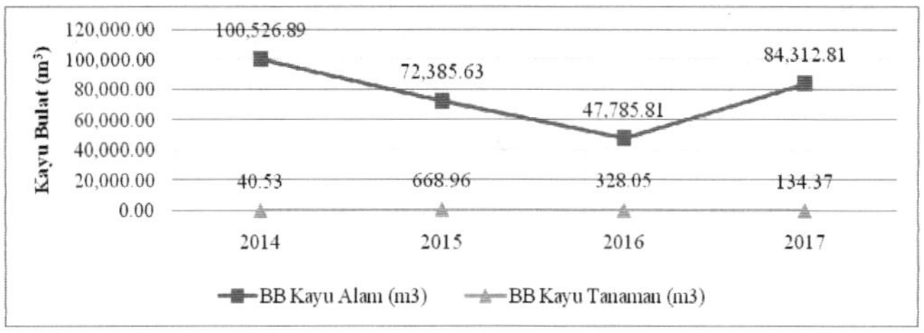


**Figure 3.** Use of logs raw material in 2014-2017 ($m^3$)

Other information flows related to marketing plywood products. The marketing activity of plywood products at this company is handled by the marketing section. The numbers of employees in the marketing department are 20 persons. The negotiation with customers in terms of the number of requests and prices was handled directly by the marketing section at the head office in Jakarta. 94.5% of plywood products intended for export market to countries such as Japan, Europe, Taiwan and the UK and the rest for local market (5.5%) intended for Jakarta and Medan. To distinguish the results of plywood products produced by the company with other products, PT. KTC uses the Indonesian KTC-UP trademark for exports market. In general, plywood products marketed by PT. KTC to exports based on Japan Agriculture Standard (JAS), International Hardwood Product Association (IHPA) standards and EURO (Europe) standards.

Figure 4 shows that in 2014, the total export of plywood products was 45.050,24 $m^3$ and continued to decline in 2016 of 18,321.16 m3. However, plywood exports increased significantly by 26.190,34 $m^3$ in

2017. In addition, for the local market, the plywood continued to decline for three years in 2014 to 2016. However, there was a significant increase in 2017 accounted 2,284.6 $m^3$ from 1,380.68 $m^3$ in 2016.

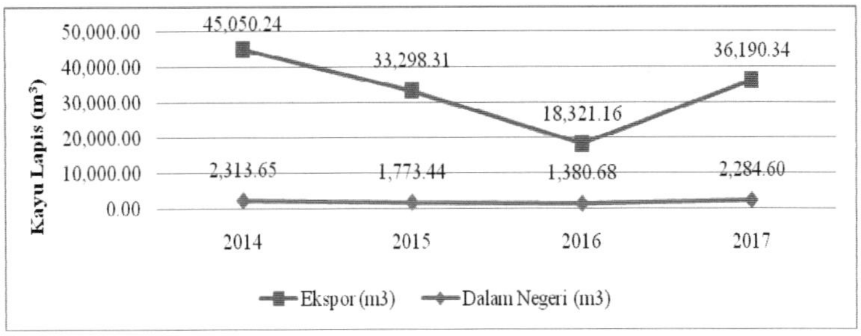


**Figure 4.** Export and local market of plywood in 2014 - 2017 ($m^3$)

Based on the data of the fulfillment of raw materials, the use of logs, and the selling of plywood, it can be seen that the fluctuations tend to decline in production and selling of plywood products. This is due to the lack of supplies of raw materials for logs. One of the factors that caused this was due to slow delivery of raw materials caused by weather factors. The largest supplier of logs is from PT. Gema Hutani Lestari in its annual period has many rainy months.

## 4. Conclusion

The flow of product flows from upstream to downstream, supplies raw materials, production processes and marketing. The source of raw materials comes from forest concessions. The financial flow is related to the system of buying and shipping plywood products, and the selling price of plywood. The system for purchasing of plywood is to use a delivery order system based on mutual agreement between producers and consumers. As for product shipping costs, PT. KTC uses the Cost and Freight system. For sale prices, PT. KTC applies different selling prices for each destination country. The flow of information flowing in the supply chain of PT. KTC is related to the system for the procurement of raw materials, the amount of raw material inventory, the amount of logs used, marketing, and how many products are exported and which are marketed locally. The system for procurement of logs is carried out under a contract system. The amount of raw material inventory until 2017 is 85,884.56 $m^3$ while the number of usage until 2017 is 36.190,34 $m^3$. For marketing activities, PT. KTC exports more plywood with 2017 exports of 36.190,34 $m^3$, while domestic is only 2.284,60 $m^3$.

## References

[1] Awaliyah M, Novitasari M, and Shantika M 2016 Analisis Produksi Kayu Lapis Menggunakan Statistical Quality Control. *Buletin Ilmiah Mat. Stat. Dan Terapannya (Bimaster)* **5** (1) pp 1–8
[2] Food and Agriculture Organization (FAO) 2009 *The State Of Food And Agriculture* (Rome, Italy: Electronic Publishing Policy and Support Branch Communication Division)
[3] Iswanto D 2013 Faktor-Faktor yang Mempengaruhi Ekspor Kayu Lapis Indonesia Ke Jepang. *Jurnal Ekonomi Pembangunan* **1** (02)
[4] Murniasih E 2008 Determinan Pertumbuhan Produktivitas Faktor Total (TFP) Industri Kayu Lapis di Indonesia. (Depok, Indonesia: Universitas Indonesia)
[5] Statistik Lingkungan Hidup dan Kehutanan 2016 Kementerian Lingkungan Hidup dan Kehutanan. Jakarta.
[6] Suryandari E Y 2008 Analisis Permintaan Kayu Bulat Industri Pengolahan Kayu (Log Demand Analysis on Forest Product Industry). *Jurnal Penelitian Sosial dan Ekonomi Kehutanan* **5** (1) pp 15–26
[7] Supit T, and Jan A H 2015 Analisis Persediaan Bahan Baku Pada Industri Mebel Di Desa Leilem. *Jurnal EMBA* **3** (1) pp 1230–1241

[8] Kurniawan 2014 Analisis Rantai Pasokan (Supply Chain) Komoditas Cabai Merah di Kabupaten Jember. *Jurnal Berkala Ilmiah Pertanian* **9** (9)

[9] Herda S, and Setyawan A A 2016 Manajemen Rantai Pasok Kayu Gaharu Di Kalimantan Barat. *Jurnal Ekonomi Manajemen Sumber Daya* **18** (2)

[10] Emhar A, Aji J M, Agustina T 2014 Analisis Rantai Pasokan (Supply Chain) Daging Sapi di Kabupaten Jember. *Jurnal Berkala Ilmiah Pertanian* **1**(3) pp 53–61

[11] Makkarennu, Nakayasu A, Osoawa K, and Ichikawa M 2013 Applied Economic Order Quqntity (EQQ) In Logs Raw Material Inventory Control For Plywood Industry. *International Journal of Agricultural Sustainability* **5** (2)

IOP Conf. Series: Materials Science and Engineering **593** (2019) 012009 doi:10.1088/1757-899X/593/1/012009

# Citric acid, an environmentally friendly adhesive and wood impregnation material-review of research

**T D Cahyono[1*], Syahidah[2]**

[1]Faculty of Forestry, Ambon Darussalam University. Jl. Waehakila Puncak Wara Ambon 97128. Indonesia.
[2]Faculty of Forestry. Hasanuddin University. Jl. Perintis Kemerdekaan Km. 10. Tamalanrea. Makassar. Indonesia.

*Email: tekatdwicahyono@gmail.com

**Abstract**. Citric acid's potency to be adhesive is proven with an increasing quality of product bonding resulted from wood and other lignosellulosic materials. Wood waste, bagasse, sorghum and corn stalks, and bamboo are quality particleboard materials after addition of 20% citric acid. Moulding made of a mixture of tree bark and citric acid with the same concentration is also able to pass repeated boiling test without having any damage. Ester bond between wood's hydroxyl group and citric acid's hydroxyl group is formed at 180°C and gets more optimal as the pressing temperature increases during production process. This results in an increase of board's physical-mechanical properties pursuant to the standard. Some researches remain presenting non-standard test parameters and this is a challenge to be answered with further research. The density of Samama wood impregnated with 10% citric acid has its density increased between 17.11% and 20.13%. Samama wood does not experience a recovery of set after it has been pressed for 50 minutes at 180°C.

## 1. Introduction

Environmental issue is one important point in future technology development, including adhesive technology for lignosellulosic materials. Alternatives to various petroleum resin based adhesives like urea formaldehyde (UF), phenol formaldehyde (PF) and melamine formaldehyde (MF) start to be searched for since they are evidently less environmentally friendly. Formaldehyde emission generated by these adhesives is hazardous to health. To reduce the emission may be made by changing adhesive formula[1], adding surface layer[2] and changing adhesive[3, 4].

Many environmentally friendly adhesive alternatives have been studied to substitute or even potentially replace formaldehyde containing materials. One of these alternatives is a compound generated from plant of genus citrus, which is citric acid. Citric acid may well bind hydroxyl groups of wood[5], thus may well be distributed and form *cross-link*. Many researches of citric acid have been presented with various methods and are continuously developed to generate the best product. In addition, citric acid has also been used as a fixation material for Samama wood densification. Find this more in Tabel 1.

**Table 1.** Citric acid as adhesive for lignosellulosic materials.

| Num. | Application | Support |
|---|---|---|
| 1 | Adhesive for Molding with tree bark material | [5-7] |
| 2 | Adhesive for particleboard with material: | |
| | a. Nypa frond and oil palm midrib | [8-10] |
| | b. Wood waste | [11-13] |
| | c. Bagasse | [14, 15] |
| | d. Sorghum and corn stalks | [16, 17] |
| | e. Bamboo | [18-21] |
| 3 | Samama wood impregnation | [22] |

## 2. Citric Acid as adhesive

### *2.1. Tree bark*

Acacia (*Acacia mangium*) tree bark powder is mixed with citric acid and poured onto iron molding at 180°C, 4 MPa for 10 minutes. The Modulus of Rupture (MOR) and Modulus of Elasticity (MOR) moldings employ adhesive of 20 wt% citric acid, each at 18.1 MPa and 4.9 GPa. The molding is not damaged during the repeated boiling treatment. Based on the result of Fourier transform infrared spectra, ester bond formation is seen between citric acid's carboxyl group and tree bark's hydroxyl group, particularly at high temperature. These chemical bonds confirm that citric acid is worthy to be used as natural adhesive[5-7].

### *2.2. Particleboard*

*2.2.1. Nypa frond (Nypa fruticans* Wurmb*) and oil palm midrib.* An addition of 10% citric acid increases the physical and mechanical properties of particleboard made of nipa palm midrib[8-10]. The optimal temperature for pressing process is 180°C with a duration of 15 minutes. In other research employing oil palm midrib, the optimal temperature for the best quality is 200°C with a duration of 10 minutes[9].

*2.2.2. Wood waste. Albizia chinensis* wood waste based particleboard is made using a mixed adhesive of citric acid and sucrose. The sucrose addition serves to provide hydroxyl groups and increase the number of ester groups. Sucrose mixture in adhesive formula evidently increases particleboard's physical properties. On the contrary, sucrose without citric acid does not show a good binding property at 180°C[11]. The optimal temperature for particleboard making from wood waste is 200°C with specific pressure of 3.6 MPa and duration of 10 minutes. The best quality wood is shown with a mixed adhesive of citric acid and sucrose 50:50 wt%, with solid content 15 wt%[12]. Other research with a ratio of adhesive to solid content 25:70 and 30 wt% shows the internal bonding (IB), MOR and thickness swelling (TS) values pursuant to JIS A5908 (2003) for board type 18[23]. The wood waste used is that of softwood with making process at 200°C and pressing duration for 10 minutes.

*2.2.3. Bagasse. Saccharum sp.* processing at sugar mill leaves bagasse. This material may be made particleboard using citric acid adhesive[14]. The best quality particleboard in the research uses 10% citric acid at press temperature 200°C. The specific pressure of the production process is 2 MPa with duration of 10 minutes. The particleboard produced has its density of 0.66 g/cm$^3$ with indicators of quality including 4.88% moisture content, 1.98% thickness swelling, 29.29% water absorption, 1121 N/mm$^2$ modulus of elasticity, 5.04 N/mm$^2$ modulus of rupture and 0.17 N/mm$^2$ internal bonding. Furthermore, other research presents a combination of 20 wt% citric acid, press temperature 200°C for 10 minutes which produces the best quality sweet sorghum (*Sorghum bicolor*) bagasse particleboard[15]. The board's Dry bending (DB), Internal bonding (IB) and Thickness swelling (TS) are pursuant to JIS A 5908 (2003). The results of infrared spectra analysis show that ester's degree of bonding increases in line with an increase of temperature and press time.

*2.2.4. Sorghum Stalk and Corn Stalk.* Particleboard produced from a mixture of sorghum stalk and *Albizia chinensis* wood waste has also been reported using citric acid adhesive. The best quality board is produced from a combination of 100% sorghum stalk with press temperature 200°C, press time 10 minutes and pressure 25 kgf/cm$^2$. Usage of 20% adhesive results in the best quality board in comparison with 15% adhesive[16]. Meanwhile, for particleboard made of corn stalk, the use of 20% citric acid adhesive also shows board with Japanese quality standard, JIS A5908 (2003)[17]. Figure 1 shows display comparison between corn stalk particleboard with citric acid and without citric acid adhesives.

*2.2.5. Bamboo.* Betung bamboo (*Dendrocalamus asper*) particle may be made good quality board using 20 wt% citric acid[18]. This board making process is performed at pressure temperature 200°C for 10 minutes, resulting in optimal IB, MOE and MOR values of 0.44 MPa, 15.1 MPa and 4.6 GPa, respectively. The board's IB and MOR values are pursuant to JIS A5908 (2003), particularly particleboard type 8. Its MOE value is pursuant to a higher grade, which is type 13. In comparison with bamboo composite board using formaldehyde based adhesive, either particleboard products, composite beam[20] or OSB[19], the results of this research further confirm the advantage of citric acid as adhesive material. Besides mechanical properties, the dimension stability of particleboard using betung bamboo, tali bamboo (*Gigantochloa apus*) and hitam bamboo (*Gigantochloa atroviolacea*) particles also increases and is pursuant to JIS A5908(2013) after citric acid addition[21]. However, the type of bamboo factor does not present significant difference in physical and mechanical properties.

(a) (b)

**Figure 1.**Particleboard using citric acid (a) and without citric acid (b). Photo: [17].

## 3. Citric Acid improves Samama wood density

*3.1. Adhesive impregnation for wood density improvement*

This paper tries to elaborate citric acid usage as adhesive, including presentation of adhesive impregnated into wood. An impregnant generally used is formaldehyde based resin [24-28]. The purpose of such quality modification is to improve wood's density, dimension stability and strength [29, 30]. Wood's durability from deterioration caused by physical, chemical and biological factors are bonus for each treatment.

Various adhesive impregnation methods have been performed, from simple to complicated ones. One of the methods is pre-compression, particularly applied to wood with low density. This method is able to improve water absorption into wood[31]. The pre-compression method is certainly not applied independently, but combined with other methods for adhesive maximum retention.

### *3.2. Characteristics of Samama impregnated with citric acid*

Researches of citric acid impregnation using pre-compression method have been conducted on Samama wood[22]. Pre-compression is performed at 100°C for 1 hour until reaching drying set condition. The second phase is to put wood into *autoclave* under pressure between 5 – 7.5 kg/cm$^2$ (at room temperature) for 4 hours, submerged in citric acid solution. The solutions used are 5% and 10%. The third phase is repressing at 180°C with press time variable of 10, 20, 30, 40, 50 and 60 minutes. This series of method has successfully improved Samama's oven-dry density at 0.44 - 0.45 g/cm$^3$, an increase between 17.11 - 20.13%. This density increase is relatively low compared to the density increase of wood impregnated using phenol formaldehyde adhesive. The reason of this is that phenolformaldehyde adhesive generally fills in lumen, while citric acid is expected to get into cell wall and form ester bond like what occurs in particleboard.

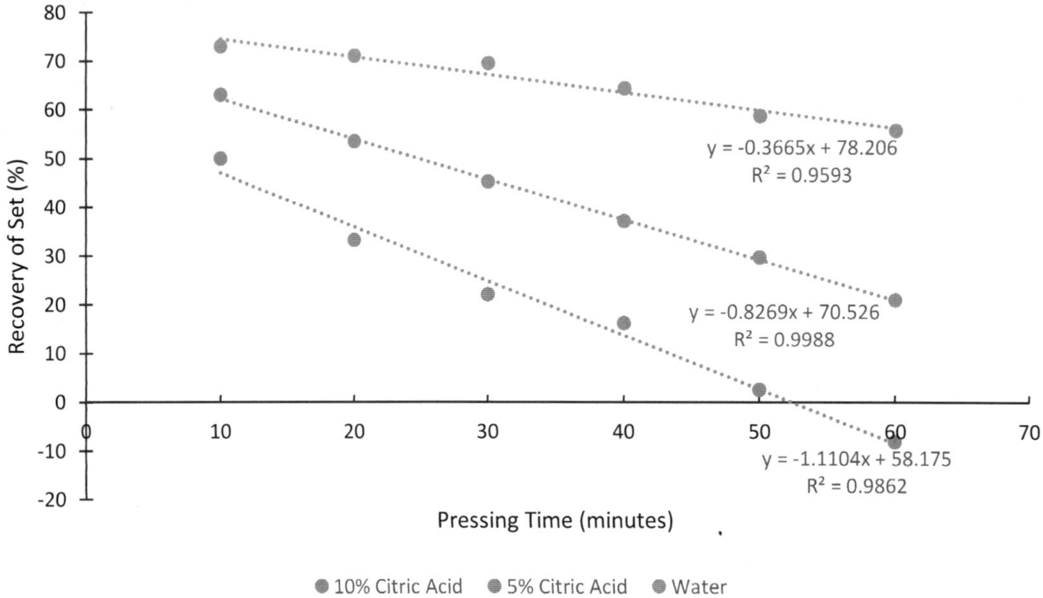


**Figure 2.** The recovery of set of wood test sample impregnated with citric acid[22].

Figure 2 shows that the recovery of set (RS) of Samama without citric acid is still high (>55%). An addition of 5% citric acid seems to be able to reduce RS, but the value is still 21% with wood impregnation at 180°C for 1 hour. Samama wood fixation occurs at a concentration of 10% citric acid with press time of 50 minutes. If desired, the temperature and citric acid concentration may be increased to reduce press time.

## 4. Conclusion

Citric acid is an environmentally friendly material evidently having advantages in its usage as adhesive for wood and materials with lignocellulose. Citric acid is able to bind OH group in wood to form ester bond, especially when the production process is performed at above 180°C. The application of 20% citric acid produces quality molding and particleboard, pursuant to the standards. Some parameters which are not pursuant to the standards are continuously anticipated and some solutions have been presented for them. Furthermore, as an impregnation material, the use of 10% citric acid is able to improve Samama wood density up to 20%. A recovery of set (RS) = 0 occurs after 50 minutes of pressing at 180°C.

**References**

[1] S Uchiyama, E Matsushima, N Kitao, H Tokunaga, M Ando, Y Otsubo 2007 Effect of natural compounds on reducing formaldehyde emission from plywood, *Atmospheric Environment* **41**(38) pp 8825–8830

[2] S Kim 2010 Control of formaldehyde and TVOC emission from wood-based flooring composites at various manufacturing processes by surface finishing, *Journal of Hazardous Materials* **176**(1-3) 14–19

[3] A H Conner 1996 Urea-formaldehyde adhesive resins, *Polymeric materials encyclopedia* **11** pp 8496–8501

[4] S Kim 2009 Environment-friendly adhesives for surface bonding of wood-based flooring using natural tannin to reduce formaldehyde and TVOC emission, *Bioresource technology* **100**(2) pp 744–748

[5] K Umemura, T Ueda, S S Munawar, S Kawai 2012 Application of citric acid as natural adhesive for wood, *Journal of Applied Polymer Science* **123**(4) pp 1991–1996

[6] K Umemura, T Ueda, S Kawai 2012 Characterization of wood-based molding bonded with citric acid, *Journal of wood science* **58**(1) pp 38–45

[7] K Umemura, T Ueda, S Kawai 2012 Effects of moulding temperature on the physical properties of wood-based moulding bonded with citric acid, *Forest Products Journal* **62**(1) pp 63–68

[8] B H Wicaksono, R Widyorini 2015 Pengaruh Perbedaan Bahan Baku dan Jumlah Asam Sitrat Terhadap Sifat Papan Partikel Dari Pelepah Nipah (*Nypa sp.*), Kehutanan, UGM, Yogyakarta

[9] K Agus, R Widyorini 2013 Pengaruh jumlah asam sitrat, suhu dan waktu pengempaan terhadap sifat papan partikel dari pelepah kelapa sawit, Kehutanan, Universitas Gadjah Mada, Yogyakarta

[10] R Widyorini, T A Prayitno, A P Yudha, B A Setiawan, B H Wicaksono 2014 Pengaruh Konsentrasi Asam Sitrat dan Suhu Pengempaan terhadap Kualitas Papan Partikel Pelepah Nipah, *Jurnal Ilmu Kehutanan* **6**(1) pp 61–70

[11] R Widyorini, P A Nugraha, M Z A Rahman, T A Prayitno 2016 Bonding ability of a new adhesive composed of citric acid-sucrose for particleboard, *BioResources* **11**(2) pp 4526–4535

[12] P A Nugraha, R Widyorini 2014 Pengaruh jumlah dan komposisi perekat alami asam sitrat-sukrosa terhadap sifat papan partikel limbah gergajian kayu sengon, Teknologi Hasil Hutan, Universitas Gadjah Mada, Yogyakarta

[13] K Umemura, O Sugihara, S Kawai 2013 Investigation of a new natural adhesive composed of citric acid and sucrose for particleboard, *Journal of wood science* **59**(3) 203–208

[14] A Damarraya, R Widyorini 2013 Pengaruh Jumlah Asam Sitrat dan Waktu Kempa Panas terhadap Sifat Papan Partikel dari Ampas Tebu, Teknologi Hasil Hutan, Universitas Gadjah Mada, Yogyakarta

[15] S S Kusumah, K Umemura, I Guswenrivo, T Yoshimura, K Kanayama 2017 Utilization of sweet sorghum bagasse and citric acid for manufacturing of particleboard II: Influences of pressing temperature and time on particleboard properties, *Journal of Wood Science* **63**(2) pp 161–172

[16] S Mardhatillah 2018 Karakteristik Papan Partikel dari Campuran Sengon dengan Batang Sorgum Menggunakan Perekat Asam Sitrat

[17] L Octaviana 2017 Kerekatan Papan Partikel Batang Jagung dengan Perekat Asam Sitrat

[18] R Widyorini, A P Yudha, Y Adifandi, K Umemura, S Kawai 2013 Characteristic of Bamboo Particleboard Bonded with Citric Acid, *Wood Research Journal* **4**(1) pp 31–35

[19] F Febrianto, W Hidayat, E S Bakar, G J Kwon, J H. Kwon, S I. Hong, N H. Kim 2012 Properties of oriented strand board made from Betung bamboo (*Dendrocalamus asper* (Schultes. f) Backer ex Heyne), *Wood science and technology* **46**(1-3) pp 53-62

[20] T D Cahyono, E Novriyanti, E T Bahtiar, M Y Massijaya 2014 Development of composite beams made from tali (Gigantochloa apus) and hitam bamboo (Gigantochloa atroviolacea), *Journal of the Indian Academy of Wood Science* **11**(2) pp 156–161

[21] R Widyorini, K Umemura, R Isnan, D R Putra, A Awaludin, T A Prayitno 2016 Manufacture and properties of citric acid-bonded particleboard made from bamboo materials, *European Journal of Wood and Wood Products* **74**(1) pp 57–65

[22] A Rumbaremata, T D Cahyono, T Darmawan, S S Kusumah, F Akbar, W Dwianto 2018 Impregnasi Kayu Samama [Anthocephalus macrophyllus (Roxb.) Havil.] Menggunakan Asam Sitrat dengan Metode Pre-kompresi untuk Meningkatkan Kerapatannya, Seminar Lignoselulosa, (Bogor, Indonesia)

[23] K Umemura, O Sugihara, S Kawai 2015 Investigation of a new natural adhesive composed of citric acid and sucrose for particleboard II: effects of board density and pressing temperature, *Journal of wood science* **61**(1) pp 40–44

[24] M I Shams, H Yano 2011 Compressive deformation of phenol formaldehyde (PF) resin-impregnated wood related to the molecular weight of resin, *Wood science and technology* **45**(1) pp 73–81

[25] R Hartono, I Wahyudi, F Febrianto, W Dwianto 2017 Maximum compression level measurement of oil palm trunk, *Jurnal Ilmu dan Teknologi Kayu Tropis* **9**(1) pp 73–83

[26] T Furuno, Y Imamura, H Kajita 2004 The modification of wood by treatment with low molecular weight phenol-formaldehyde resin: a properties enhancement with neutralized phenolic-resin and resin penetration into wood cell walls, *Wood Science and Technology* **37**(5) pp 349–361

[27] E Bakar, P Tahir, M Sahri, M Mohd Noor, F Zulkifli 2013 Properties of resin impregnated oil palm wood (Elaeis guineensis Jack), *Pertanika J. Trop. Agric. Sci* **36** pp 93–100

[28] M Inoue, S Ogata, S Kawai, R M Rowell, M. Norimoto 2007 Fixation of compressed wood using melamine-formaldehyde resin, *Wood and fiber science* **25**(4) pp 404–410

[29] S Kumar 2007 Chemical modification of wood, *Wood and Fiber Science* **26**(2) pp 270–280

[30] J E Jakes, C G Hunt, D J Yelle, L Lorenz, K Hirth, S C Gleber, S Vogt, W Grigsby, C.R. Frihart 2015 Synchrotron-based X-ray Fluorescence Microscopy in Conjunction with Nanoindentation to Study Molecular-Scale Interactions of Phenol–Formaldehyde in Wood Cell Walls, *ACS applied materials & interfaces* **7**(12) pp 6584–6589

[31] L Suryanegara, W Dwianto 2004 Pengaruh Dimensi dan Pre-kompresi Kayu terhadap Sifat Penyerapan Air (Dimensions and Pre-Compression Wood Influences on Water Absorption), *Jurnal Ilmu dan Teknologi Kayu Tropis* **2**(2) pp 79–82

IOP Conf. Series: Materials Science and Engineering **593** (2019) 012010 doi:10.1088/1757-899X/593/1/012010

# Resinless Plywood Production by Using Oxidized Acacia Bark Powder as a Binder

**S Suhasman[1*], Agussalim[1]**

[1]Laboratory of Forest Product Utilization and Processing, Faculty of Forestry, Hasanuddin University, Makassar, Indonesia

*Email: suhasman@yahoo.com

**Abstract**. Nowadays, the established technology in plywood industries for producing plywood is still using adhesive as veneer bonding agent. Utilization of the adhesive has several of weaknesses such as containing formaldehyde compound and using the unrenewable raw material. The purpose of this research was to develop a novel technique to produce plywood without involving adhesive as a binder. The oxidized bark powder of *Acacia mangium* was utilized as a binder to substitute the adhesive. Bark powder was oxidized by using hydrogen peroxide on four levels (5%, 10%, 15%, 20%) and the catalyst. The result showed that utilization of oxidized bark powder can be applied in producing resinless plywood. Unless the plywood with 15% of hydrogen peroxide, the other plywood fulfil the National Indonesian Standard (SNI) No 01-5008.2-2000 where min. 7 kg/cm$^2$ as the requirement. The plywood with 20 percent of hydrogen peroxide has the highest bonding strength (13.24 kg/cm$^2$). After immersing this plywood in hot water (70°C) for 2 hours and drying in the oven (60°C) for 3 hours, the wood failure test also showed a good value, namely 76.15 percent. These facts indicated that the bonding and the water resistant of the plywood were relatively high. It tends that the technology for producing resinless plywood is potential to be developed.

## 1. Introduction

In 2017, Indonesia's plywood industry manufactured 3.2 million m3 of plywood products [1]. The export value of this product reached 31.87 trillion rupiahs. This value was the third largest of wood products after pulp and paper. In addition to being the important commodity of forest product, utilization of plywood is getting widespread Indonesia and gradually replacing solid-wood domination.

The industries have been using formaldehyde-based adhesives in the manufacturing of plywood. It is an issue because the resins are known to produce formaldehyde emission that is harmful to health and can trigger cancer [2], mainly if plywood is used in a limited ventilation room [3]. Some additives were given to reduce the emission such as urea [4,5,6], ammonium carbonate [4,6], melamine [5,6], ammonium hydroxide [3], and activated charcoal [7,8], but the materials have not been able to reduce the emission totally. The use of urea and ammonium carbonate resulted in a decrease the bonding strength.

This study is designed as a new approach for producing high-grade plywood that is much more environmentally friendly, namely technology of plywood manufacturing without using adhesives or resinless plywood. This approach is adopted from the technology that has been intensively developed in producing particleboard, but it is unproven its possibility in plywood manufacturing technology. The basic principle in this technology is the activation of lignin or tannin that are capable of functioning as a binding agent.

The veneers on plywood have low accessibility by oxidizing materials due to its solid form; therefore, it needs an additional ingredient on adhesive line that can increase accessibility of the chemical components and bonding strength. In this study, using bark powder of acacia which has high tannin and relatively easily oxidized may petrify in stronger bonding formation between the veneers.

## 2. Method

This research was conducted by using sengon (*Paraserianthes falcataria*) as veneer material, and bark of acacia (*Acacia mangium*) as bonding agents. Sengon veneer was obtained from Bogor, and acacia bark was collected from Sinjai. The veneers were air-dried to reach 12% ± 2 % of moisture contents. The bark was made to be particles with 20 to 40 mesh of dimension.

Plywood was made with applying four oxidation treatments, they were 5, 10, 15 and 20 percent hydrogen peroxide based on oven-dry particle weight, and all of them used 5 percent of the catalyst based on hydrogen peroxide weight. Oxidation process refers to Suhasman et al. Method [9]. Weight of the particles used on each adhesive line was 200 g/m$^2$. The 30 cm by 30 cm by 0.3 cm plywood was manufactured with hot pressing at a pressure of 10 kgf/cm$^2$ and temperature of 180°C for 5 minutes. Plywood was then conditioned for two weeks. Physical and mechanical tests were determined based on Indonesian National Standard 01 5008.2 2000.

Tannin content was determined by soaking 40 to 60 mesh of wood bark powder on 96% ethanol for three days. The extraction filtrate was then made more viscous by using rotary evaporator. Furthermore, the viscous extractive was analyzed the tannin quantitative by using UV spectrophotometer.

## 3. Result and Discussions

Figure 1 shows density averages of resinless plywood using acacia bark powder as bonding agents on four levels of hydrogen peroxide. The density values represent that the plywood with 20% of H2O2 has the highest density (0.6 g/cm3). Meanwhile, the plywood with 5% of H2O2 is the lowest (0.52 g/cm3). However, the result shows the plywood has a slight difference in density. It can be caused by a similar thickness of veneers used. In general, there is not significantly different density among the plywood.

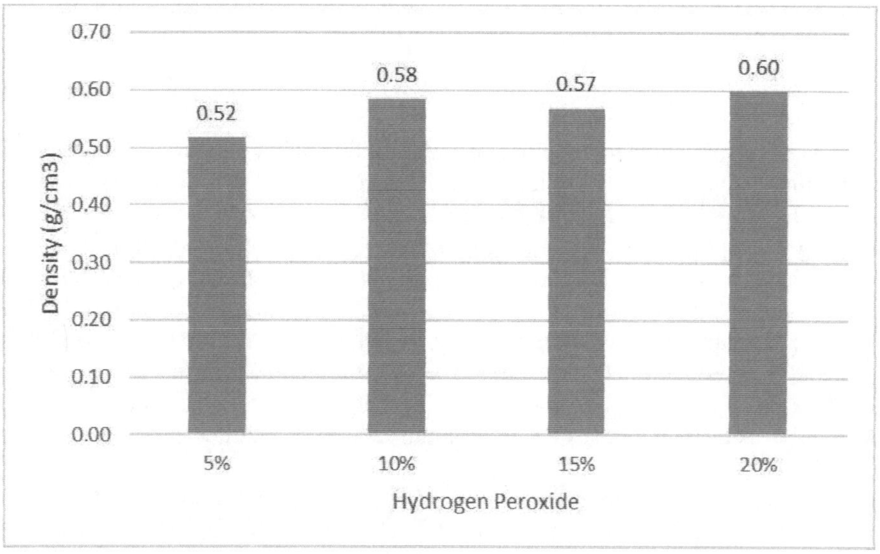


**Figure 1.** Resinless Plywood denstity

Moisture content (MC) of the plywood are showed in Figure 2. The plywood with 15 and 20 percent of hydrogen peroxide have moisture content lower than 6 percent. The plywood with 5 percent of hydrogen peroxide (6.84 percent) is the highest MC. However, the moisture content of the plywood is not significantly indifferent, and they fulfil the SNI (14%). In general, the moisture content of plywood is a quite low average of 6 percent. It possible happens because there is densification of the veneer.

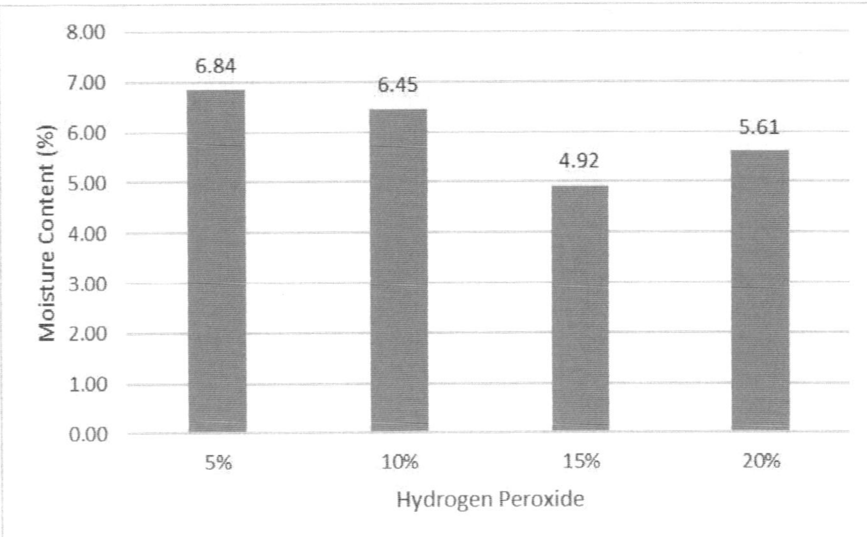


**Figure 2.** Resinless Plywood Moisture Content

As shown in Figure 3, the plywood with 20 percent of hydrogen peroxide has the highest bonding strength (13.24 kg/cm$^2$), and it is almost two times higher than other. It also has a great value of wood failure (76.15 percent) based on testing result. In general, only the plywood with 15 percent of hydrogen peroxide does not fulfil the SNI, but the value is slightly different to the plywood with 5 and 10 percent of hydrogen peroxide where both meet to the standard.

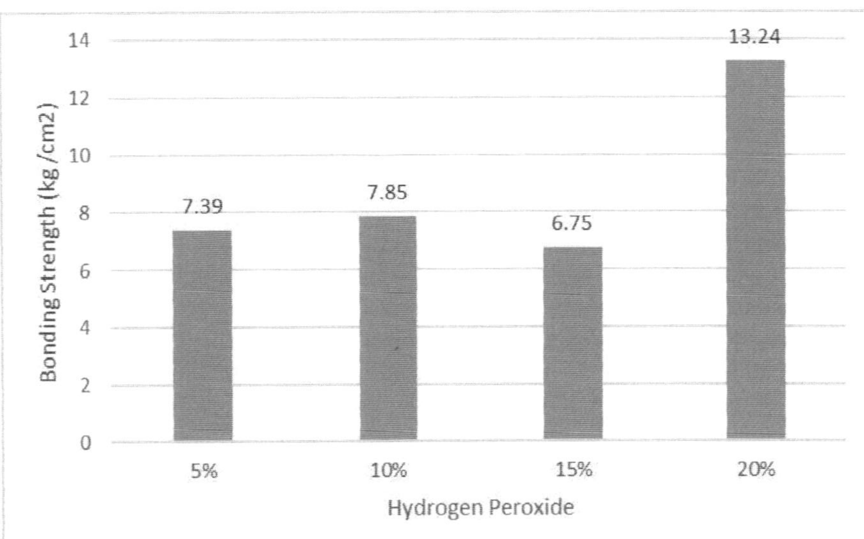


**Figure 3.** Bonding strength of resinless plywood

Suhasman [10] reported that internal bond of sengon binderless particleboard with oxidation method was affected by the number of hydrogen peroxide which the higher number of $H_2O_2$ the lower their internal bond. It is caused by a high reactivity of this species of the oxidant, so they do not need to use $H_2O_2$ in large number. However, it is different for resinless plywood in which using 5 to 15 percent of hydrogen peroxide show the relatively same bonding strength and increasing significantly on 20 percent.

The wood failure of the plywood with 20 percent of hydrogen peroxide showed an excellent characteristic. Meanwhile, the bonding strength is also affected by tannin content of the bark. Based on analyses resulted, the bark powder of acacia contained 6.2 percent of tannin. The relation between bonding strength and tannin content is caused by phenolic compound of tannin, especially hydrolysable tannins in which they have hydroxyl groups that are potentially oxidized and form radical groups.

## 4. Conclusion

Oxidation treatment to produce Resinless Plywood is possible to be applied. The results showed that utilization of bark powder of acacia with 20 percent of hydrogen peroxide as a bonding agent is the highest bonding strength (13.24 kg/cm2) which the value is two times larger than other and it has a good value of wood failure (76.15 percent). Only plywood with 15 percent of hydrogen peroxide does not fulfil SNI for bonding strength (> 7 kg/cm2).

## References

[1] Ministry of Environment and Forestry of the Republic of Indonesia 2018 Statistik Lingkungan Hidup dan Kehutanan Tahun 2017 (Jakarta: Kementrian Lingkungan Hidup dan Kehutanan) p 188

[2] Li K 2002 Use of Marine Adhesive Protein as a Model to Develop Formaldehyde-Free Wood Adhesives (Oregon: Proceeding the 6th Pacific Rim Bio-Based Composites Symposium)

[3] Santoso A and Sutigno P 2004 Pengaruh Fumigasi Amonium Hidroksida Terhadap Emisi Formaldehida Kayu Lapis dan Papan Partikel *Jurnal Penelitian Hasil Hutan* **22**(1) pp 9–16

[4] Sutigno P and Santoso A 1990 Pengaruh Penambahan Urea dan Amonium Karbonat pada Perekat Urea Folmadehida terhadap Emisi Formaldehida dan Keteguhan Rekat Tripleks Jabon *Jurnal Penelitian Hasil Hutan* **7**(1) pp 12–16

[5] Sutigno P and Santoso A 1995 Pengaruh Penambahan Urea dan Melamin pada Perekat Urea Folmadehida terhadap Emisi Formaldehida dan Keteguhan Rekat Tripleks Meranti *Jurnal Penelitian Hasil Hutan* **13**(4) pp 144-152.

[6] Santoso A and Sutigno P 1998 Several Factors Affecting the Formaldehyde Emission from Wood-based Panels (Bogor: Proceedings The fourth pacific rim bio-based composites symposium)

[7] Pari G, Sofyan K, Syafi'I W, and Buchari 2004 Arang Aktif Sebagai Bahan Penangkap Formaldehida Pada Kayu Lapis *J. Tek. Ind. Pert.* **14**(1) pp 17–23

[8] Santoso A and Pari G 2015 Sifat Papan Partikel Daur Ulang Rendah Emisi Formaldehida. *Jurnal Penelitian Hasil Hutan* **33**(1) 1–10

[9] Suhasman, Massijaya M Y, Hadi Y S, and Santoso A 2011 Particle Oxidation Time for the Manufacture of Binderless Particle Board *Wood Research Journal* **2**(1) 9–18

[10] Suhasman, Massijaya M Y, and Saad S 2011 Physical and Mechanical Properties of Binderless Particleboard Made from Candlenut Wood in Various Oxidator Levels (Yogyakarta: Proceedings The 3rd International Symposium of Indonesian Wood Research Society)

## Acknowledgement

We acknowledge the sponsorship of the Hasanuddin University for conducting this research through excellent research of Hasanuddin University research grant scheme 2017.

# Study of Bee Forage Based on Bee Bread From 4 Species of Honeybees in Teaching Forest Hasanuddin University

**A Sadapotto[1], Juniastuti[1], Budiaman[1]**

[1]Forest Department, Forestry Faculty, Hasanuddin University, Jl. Perintis Kemerdekaan, Makassar, Indonesia

**Abstract.** This study was intend to know and to identify bee forage in bee bread from honeybee in Teaching forest Hasanuddin University. Study method used in this study was survey method and identification of plant species were done in Plantation Industry Laboratory. There were 2 stage of the study i.e. collecting sample of bee bread from 4 species of honeybee (*Apis cerana, Apis mellifera, Apis dorsata and Trigona sp*) and identification of pollen in Plantation Industry Laboratory. Study results indicate that there were 20 pollen species i.e. banana (*Musa paradisiaca),* papaya (*Carica papaya*) , Mango (*Mangifera indica)* puteri malu (*Mimosa pudica*), eggplant (*Solanum tovarum*), sangitan (*Sambucus javanica),* water apple (*Syzgium aqueum),* candlenut (*Aleurites moluccana*), jatropha (*Jatropha curcas*), euphorbia (*Ephorbia milii*), guava (*Psidium guajava*), Acacia (*Acacia mangium),* sugar palm (*Arenga pinnata*), nusa indah (*Mussaenda pubescens*), chili (*Capsicum annum*), puspa (*Schima wallichii*), coconut (*Cocos nucifera*), corn (*Zea mays),* jackfruit (*Artocarpus heterophyllus*), and Piperaceae. The highesr bee bread from 4 species of honeybee pollen was sugarpalm (*Arenga pinnata)* with percentage 15,08%. The lowest bee bread was guava (*Psidium guajava)* with percentage 0,09%. The most pollen collecting was *Trigona sp* with 10 plant species.

## 1. Introduction

Honeybee is an insect producing honey that since long time ago have been known. Since long time ago human hunt honeybee in the cave, the tree hole and other place to take the honey. Honeybee produce the product need of medical world like royal jelly, pollen, bee bread, bee wax and others. Bee forage source consist of nectar and pollen that originated from plant flower.

Bee bread is the one of honeybee product that have multiple use as food and medicine. Bee bread is bee pollen that made naturally by honeybee, collected from male flower mixed with nectar that sucked and honeybee saliva which kept in the pocket between the two legs.

Teaching forest of Hasanuddin University consist of various plant species, fauna, and useful insect. Useful insect that present in Teaching Forest are honey bees. Based on field survey, honeybee found in Teaching Forest is *Trigona* sp, *Apis cerana*, *Apis dorsata*,, *Apis mellifera*. The succcesful of honeybee production depend on bee forage source that exist. Plant species used by honeybee as bee forage source was various. From many kind of plant that used as bee forage source in Teaching Forest, some of them potentially to produce bee bread and honey bee.

This study was intended to know and identify plant species that exist in Bee bread in Teaching Forest Area. The usefullness of this study was to give information about bee forage source that prefer by honeybee that exist in teaching forest Hasanuddin University related to reforestation plant species choice.

## 2. Study Method

*2.1.Time and Location of The Study*

Sample taking of bee bread was done in the area of Forest Teaching Hasanuddin University Maros Regency. The study was held from January to March 2016. Identification of bee forage was done il Laboratory of Industry Plantation Office, Makassar.

*2.2 Equipment and Material*

Equipment used in this study were writing tools, label, tally sheet, heater, microscope, deck glass, object glass, stryrer, test tube, spatula, pinset, drop pipet, camera. Material used was bee bread from each colony, bee wax, and chemical that consist of 70% gliserin and safranin.

*2.3 Study Procedure*

a. To take bee brood sample 5 mg from honeybee colony Trigona sp, Apis dorsata, Apis cerana, Apis mellifera
b. To put each of the bee bread in sealed container.
c. To dissolve the 2 mg of bee bread with 2 ml alcohol 70% in the container, to add 1 ml gliserin and 1 ml safranin in 10 minutes.
d. To stir slowly and the to put a little bit the sample with pipet and then to put the sample in object glass.
e. To cut the bee wax in small pieces, and to put in the edge of preparation, to heat until bee wax molten and the close it with deck glass.
f. To observe under the microscope with 100 time magnification
g. To identify the origin of plant species pollen by using pollen identification book

*2.4.Variable observed*

1. Origin of plant pollen species
2. Amount of plant pollen species in the preparation
3. Amount of pollen type each honeybee species

*2.5.Data analysis*

1. Amount of pollen in each bee bread
2. Percentage of pollen in each plant in bee bread with the formula :

$$X = \frac{\sum A}{\sum B} \times 100\%$$

Keterangan X = percentage of pollen amount in each plant in bee bread (100%)
A = amount of pollen in each plant in the preparation
B = amount of pollen in all plant in the preparation

IOP Conf. Series: Materials Science and Engineering **593** (2019) 012011 doi:10.1088/1757-899X/593/1/012011

## 3. Result and Discussion

### *3.1. Average percentage of the amount of Pollen in Bee Bread*

#### *3.1.1 Trigona sp*

Based on the study we can get the result in Fig. 1.

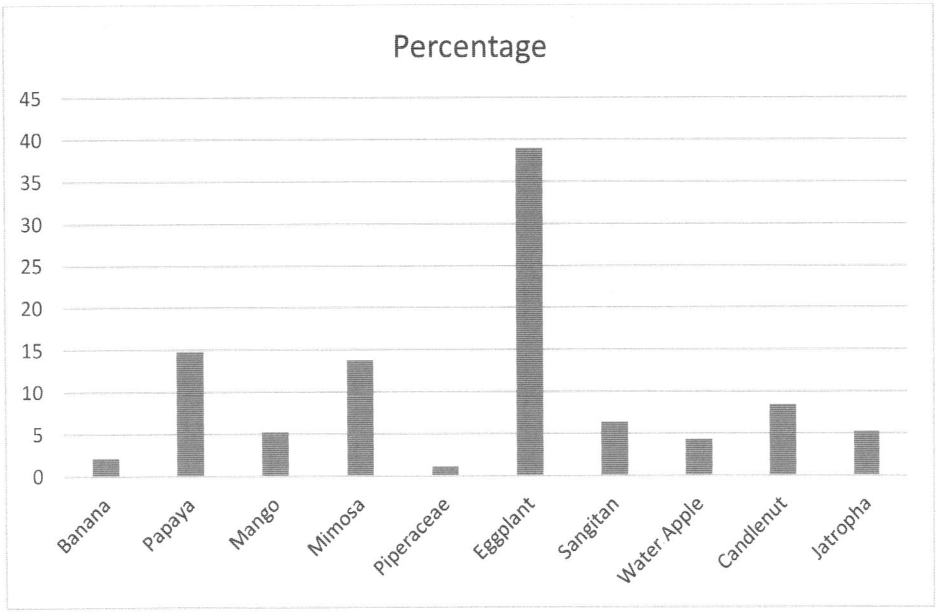


**Figure. 1.** Average percentage of amount of pollen in the bee bread of Trigona sp honeybee

In Figure. 1 we can see source of pollen and bee bread of Trigona sp were banana, papaya, mango, Mimosa, Piperaceae, eggplant, sangitan, water apple, candlenut, jatropha. Plant species most preferred by Trigona sp was eggplant with precentage 38,95% while the less preferred was Piperaceae with percentage 1,05%.

Eggplant dominated because when honeybee collected pollen source coincide with flowering season and distance between honeybee colony and pollen source (eggplant) about 7 meter.

#### *3.1.2. Apis dorsata*

In Fig. 2 indicated that source of pollen from bee bread of *A. dorsata* honeybee was corn, mango, acacia, jatropha, sugar palm, jackfruit. Bee forage plant most preferred was sugar palm with percentage 36,96% while the less preferred was acacia with percentage 2,7 %.

Identification of pollen indicated that most collected pollen by A dorsata honeybee was pollen from sugarpalm. Sugarpalm dominated because when honyebee collected bee forage coincide with flowering season and distance between honeybee colony and sugarpalm about 8 metre.

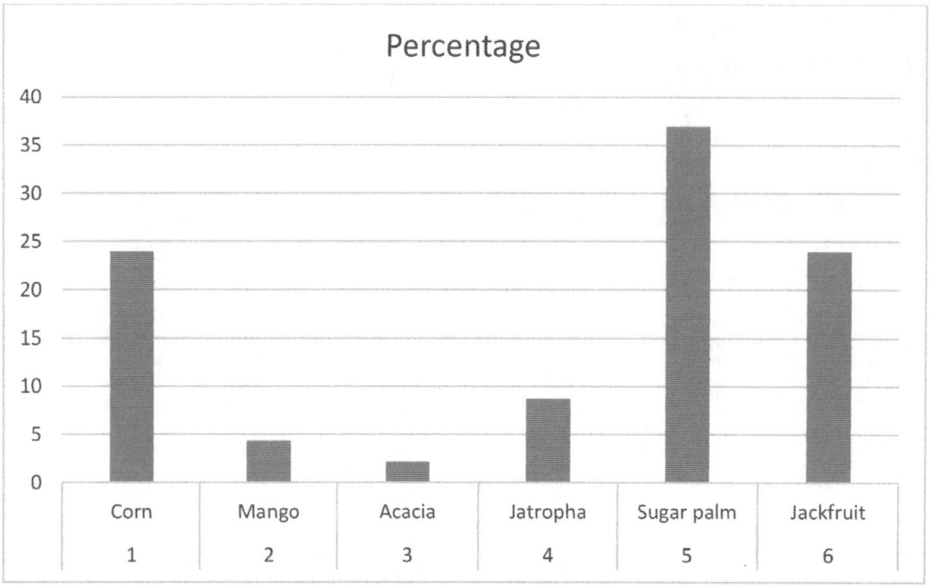


**Figure. 2**. Average percentage of amount of pollen in the bee bread of *Apis dorsata* honeybee

*3.1.3. Apis cerana*

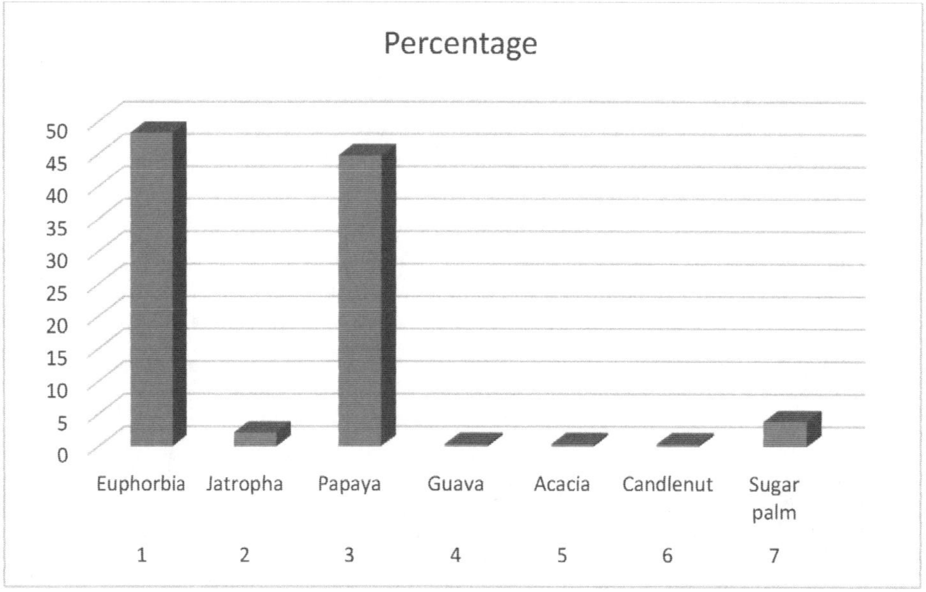


**Figure 3**. Average percentage of amount of pollen in the bee bread of *Apis cerana* honeybee

In Figure. 3 we can see the source of pollen from bee bread in A. cerana colony were Ephorbia (*Euphorbia milii*), Jatropha (*Jatropha curcas*), Papaya (Carica papaya), Guava (*Psidium guajava),* Acacia (*Acacia mangium*), candlenut (*Aleuritus moluccanus*), sugar palm (*A. pinnata)*. Bee forage that most preferred by A. cerana honeybee was euphorbia with percentage 48,25%, while the less preferred was acacia (*Acacia mangium*), candlenut (*Aleuritus moluccana)* and guava (*Psidium guajava*).

Identification of pollen indicated that most pollen collected by *A. cerana* honeybee was pollen from Ephorbia (*Euphorbia milii*). Pollen from Ephorbia dominated because honeybee collected the pollen coincide with flowering season and the distance between the honeybee colony and euphorbia was 6 metre.

*3.1.4. Apis mellifera*

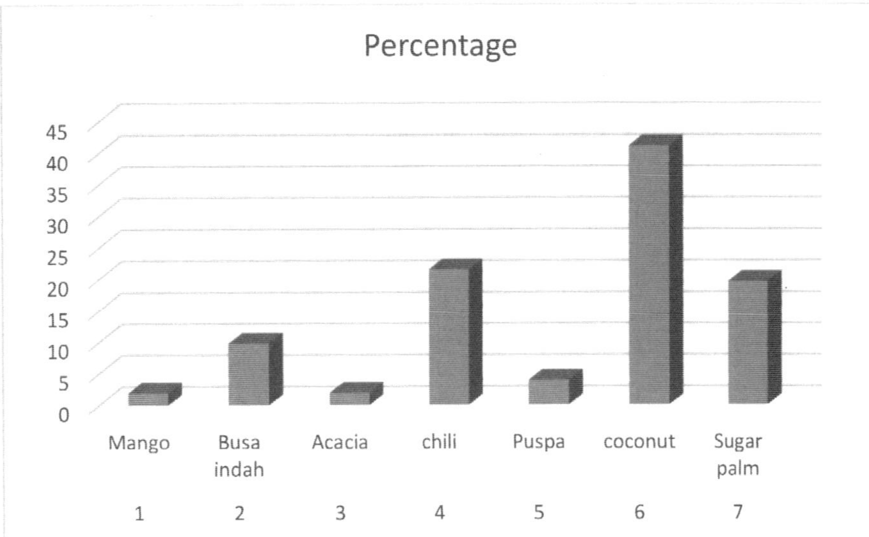


**Figure 4.** Average percentage of amount of pollen in the bee bread of *Apis cerana* honeybee

In Figure 4. We can see that pollen bee forage from bee bread of *A. mellifera* were mango (*Mangifera indica*), Nusa indah (*Mussaenda pubescens*), acacia (*Acacia mangium*), chili (*Capsicum annum*), puspa (*Schima wallichii*), coconut (*Cocos nucifera*), sugar palm (*A. pinnata*). The most preferred plant by *A. mellifera* honeybee were coconut (*Cocos nucifera*), sugar palm (*A. pinnata*). The most preferred bee forage by *A. mellifera* honeybee was coconut (*Cocos nucifera*) with percentage 41,18%, while the less preferred was mango (*Mangifera indica*) with perecentage 1,96%.

Identification of pollen indicated that the most collected pollen was pollen from coconut (*Cocos nucifera*). Coconut dominated because when honeybee collected the bee forage coincide with flowering season while the distance between honeybee colony and the coconut was 50 meter.

3.2. *Average Percentage of Amount Pollen in Bee Bread from 4 species of honeybee*

Based on observation of Bee Bread sample from 4 species of honeybee in teaching forest Hasanuddin University, we found plant species of pollen indicated in Figure 5.

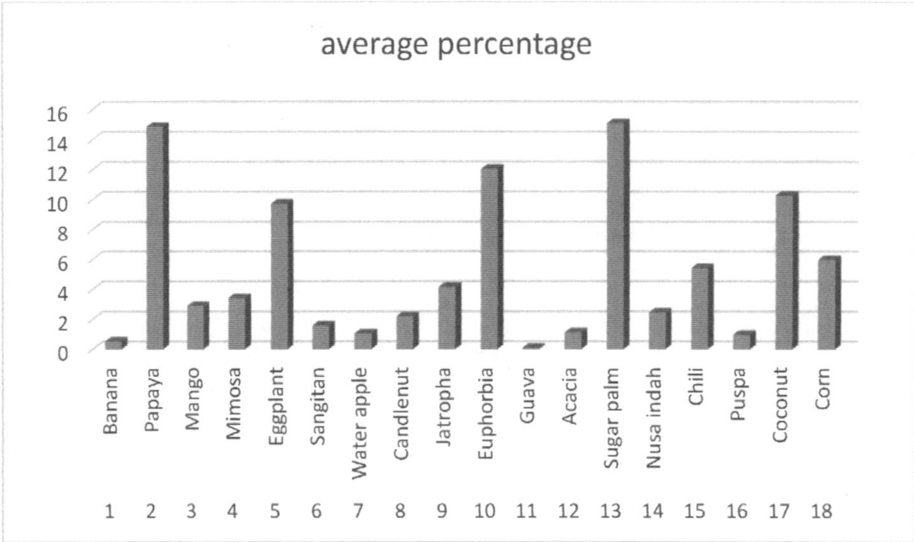


**Figure. 5**. Avarege Percentage of Amount Pollen in Bee Bread from 4 species of Honey bee (*Trigona sp., A. dorsata, A. cerana, A. mellifer*a)

In Figure. 5 we can see that there are 20 species of pollen found in bee bread of 4 species of honeybee i.e. : banana (*Musa paradisiaca),* papaya (*Carica papaya*) , Mango (*Mangifera indica)* puteri malu (*Mimosa pudica*), eggplant (*Solanum tovarum*), sangitan (*Sambucus javanica),* water apple (*Syzgium aqueum*), candlenut (*Aleurites moluccana*), jatropha (*Jatropha curcas*), euphorbia (*Ephorbia milii*), guava (*Psidium guajava*), Acacia (*Acacia mangium)*, sugar palm (*Arenga pinnata*), nusa indah (*Mussaenda pubescens*), chili (*Capsicum annum*), puspa (*Schima wallichii*), coconut (*Cocos nucifera*), corn (*Zea mays)*, jackfruit (*Artocarpus heterophyllus*), and Piperaceae.

The highest pollen sample of bee bread was sugar palm (A. pinnata) with percentage 15,08% as bee forage source of *A. dorsata, A. cerana, A. mellifer*a. Sugarpalm (*Arenga pinnata*) was the most preferrable plant by honeybee. Beside producing pollen, sugarpalm also produce nectar that was apt to honeybee [4]. Based on pollen identification in Trigona sp honeybee, it was not found sugarpalm (*A. pinnata*), because body dimension and flying distance of Trigona sp was less compared to *A. cerana, A. dorsata and A. mellifera.*

Pollen type found in bee bread was various in term of form and dimension. This indicate that honeybee take the pollen not only based on the distance to the colony, but also dimension, flavour, color of the flower.

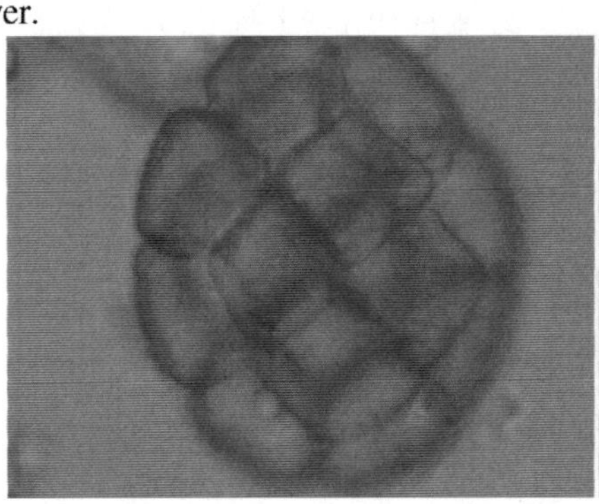
**Figure 6a.** Pollen of *Acacia mangium*

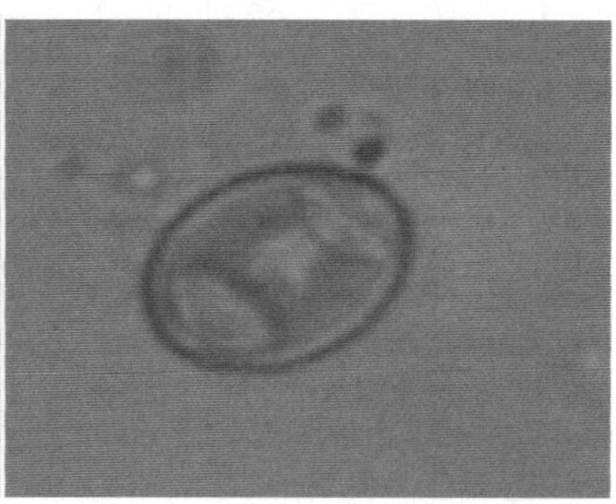
**Figure. 6b**. Pollen of *Euphorbia milii*

IOP Conf. Series: Materials Science and Engineering **593** (2019) 012011 doi:10.1088/1757-899X/593/1/012011

Honeybee preferred pollen that produced in abundancecompared to flower that produced less pollen. In normal condition, honeybee will limit to find nectar and pollen on one plant species on one visiting [1]. The role of Environet factor to pollen collection intensity held direct and indirect. Direct influence by flying activity, consumption pattern, while indirect influence by pollen production. Environment temperature influence the amount of bee forage consumption, so it will influence bee forage collection intensity from the field [2]

Honeybee collect the pollen with unique manner.i.e. descent upon flower and attach the body on flower surface so auomatically pollen attach to the honeybee body. When in break or flying, pollen collected by using pollen brush and then put into pollen basket located in rear leg. Worker bee then get into the hive and inspect the cempty cell to fiill with the pollen [4].

When pollen release in the hive comb, pollen will mix with nectar that secrete by bee worker, honey and secretion from bee workerr gland. This mixture will put into the cell by bee worker which called bee bread. The composition of bee bread different with pollen. Bee bread have higher nutrient than fresh pollen. Bee bread have lower pH and lower flour [4]. Berside that, bee bread also fermented and enzymatic reaction with the peresence of bacteria [5].

The content of bee bread i.e: iron to prevent anemia, nucleat acid to prevent aging, cellulosa to prevent constipacy, antibiotic to inhibit bacteria activity, hormon for growing.

Bee bread contain 22 kind essential amino acid, essential fatty acid, various vitamin and mineral, various enzym and hormones needed by the human body for regeneration process of tissue cel, reproduction cell, various alkaloid that have special use in stabilization of cell metabolism and growing.

### *3.3. Amount of Plant Species in Pollen from Bee Bread of 4 honeybee species*

Amount of plant species in pollen from bee bread of 4 honeybee species describe in Figure. 7

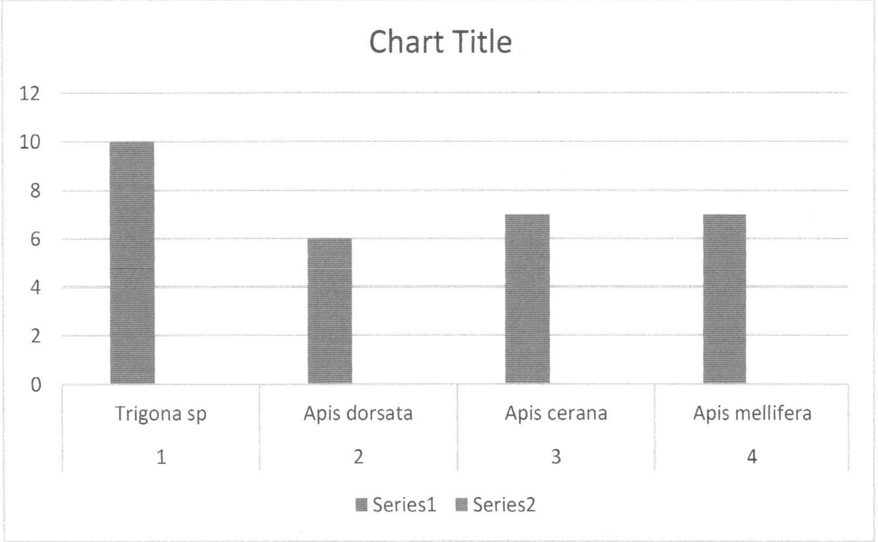


**Figure. 7.** Amount of plant species in pollen from bee bread of 4 honeybee species

In Figure 7. Indicate that the amount of plant species from pollen of 4 species honeybee bee bread, dominated by Trigona sp with 10 plant species, this because the amount of Trigona sp honeybee colony have more than the three other honeybee species and it preferred lower level flowering plant. Flying distance of Trigona sp have less radius than the three other honeybee species. One of the factor that Trigona sp get more pollen because the distance between bee forage and the colony was not far.

One of the study said that Trigona sp tend to collect more pollen than other honeybee species because the high concentration and demand of Trigona sp to bee pollen. The high concentration of Trigona sp honeybee to collect pollen because Trigona sp doesn't produce royal jelly like other honeybee [6].

## 4. Conclusion and Suggestion

### *4.1. Conclusion*

1. Plant species that source of bee forage pollen from 4 species honeybee were 20 kind of pollen i.e banana (*Musa paradisiaca),* papaya (*Carica papaya*) , Mango (*Mangifera indica)* puteri malu (*Mimosa pudica*), eggplant (*Solanum tovarum*), sangitan (*Sambucus javanica),* water apple (*Syzgium aqueum*), candlenut (*Aleurites moluccana*), jatropha (*Jatropha curcas*), euphorbia (*Ephorbia milii*), guava (*Psidium guajava*), Acacia (*Acacia mangium)*, sugar palm (*Arenga pinnata*), nusa indah (*Mussaenda pubescens*), chili (*Capsicum annum*), puspa (*Schima wallichii*), coconut (*Cocos nucifera*), corn (*Zea mays*), jackfruit (*Artocarpus heterophyllus*), and Piperaceae.
2. The highesr bee bread from 4 species of honeybee pollen was sugarpalm (*Arenga pinnata)* with percentage 15,08%. The lowest bee bread was guava (*Psidium guajava)* with percentage 0,09%.
3. The most pollen collecting was *Trigona sp* with 10 plant species.

### *4.2. Suggestion*

It need continued study in flowering season and identification of bee forage source in honey.

## References

[1] Amir, M ; Pujiastuti dan Sudarman, 1986. *Pengaruh bentuk dan warna bunga terhadap daya tarik lebah madu*. Prosiding Lokakarya Pembudidayaan Lebah Madu untuk Peningkatan Kesejahtraan Masyarakat. Perum Perhutani, Jakarta.

[2] Sihombing D.T.H. 2005. *Ilmu Ternak Lebah Madu*. Yogyakarta: Gadjah Mada University Press.

[3] Erwin. 2013. *Manfaat Bee Pollen*. http://agenmadumurni.blogspot.com/2012 /07/ manfaat-bee-pollen.html. Artikel. Diakses tanggal 2 juni 2016.

[4] Brodschneider, R. & K. Crailsheim, 2010. *Nutrition and health in honey bees. Apidologie*. Dade, H.A., 1985. Anatomy and Disection of Honey Bee. International Bee Research Association, London.

[5] Huang, Z. 2011. *Honey Bee Nutrition*. http://www. Extention. Org peges/28844/honey-bee-nutrition.

[6] Free JB. 1982. *Bees and Mankind*. Oxford London and Northampton: Alden press.

IOP Conf. Series: Materials Science and Engineering **593** (2019) 012012 doi:10.1088/1757-899X/593/1/012012

# Effect of Layer Structure on Physical and Mechanical Properties of Binderless Composite Plywood

**S Saad[1*], A D Yunianti[1], S Suhasman[1]**

[1]Laboratory of Forest Product Utilization and Processing,
Faculty of Forestry, Hasanuddin University, Makassar, Indonesia

*Email : sahriyanti_saad@unhas.ac.id

**Abstract.** Binderless composite plywood has been developed to produce formaldehyde-free composite product. Our previous research has shown that the application of oxidation treatment can be applied in the manufacture of composite plywood without adhesive. Presence of veneer in solid sheet are potentially to complicate in the oxidation process due to the reaction was very affected by accessibility of chemical component of raw materials. The main objective of this study was to determine characteristic of binderless composite plywood of Sengon wood. Variations in the physical and mechanichal properties with respect to veneer layer construction and oxidant composition of veneer and particle were analyzed.. Composite plywood were produced consist of various layers, they were 3, 5, and 7 layers means each type using 2, 3, and 4 veneers layers respectively. There were four oxidant composition between veneer and particles, namely 1 : 3, 1 : 4, 1 : 5 and 1 : 6. The results showed that the differences of composite plywood structure layers affect the characteristics of produced composite plywood. 7-layer structure and 1 : 3 of oxidant composition was the optimum combination to produce composite plywood with the best characteristics.

## 1. Introduction

Formaldehyde emission has been an issue for the particleboard industry because it could cause hazard to human health including sensory irritation and risk of developing cancer [1]. There have been many studies to reduce the risk of the emission including find formaldehyde-free adhesive such as natural resin as well as binderless particleboard. Binderless particleboard technology is one of the attractive ways to overcome emission problems of particleboard. Some researchers have been intensively developing various methods of binderless production as well as utilizing various wood or lignocellulose as raw materials. Those methods including; steam injection [2] or enzimatyc activation of particle [3] . Development of direct hot pressing method using oil palm have been reported by several researchers [4-6].Another promising method to produce binderless particleboard is oxidation method to activate particle [7]. Binderless produced board by oxidation method has several advantages such as high dimensional stability, modulus of elasticity, and internal bond. However, the modulus of rupture relatively low.

The most common work to overcome problems related to low modulus of rupture is applying veneer layer on both side of particleboard surfaces which is called composite plywood. The modulus of rupture of composite plywood may increases due to the presence of veneer layers. The outer layers of composite plywood will receive highest tensile load when panel receiving bending load. The existence of thin layer of veneer on the surface will increase the boards ability to withstand the bending loads significantly.

This study was focussed on developing binderless composite plywood products using oxidation method. Nonetheless, the feasibility of oxidation method application in producing composite plywood remain questioned. Oxidation process need to access the chemical component of wood, so there is no guarantee that solid shape material such as veneer can be oxidized successfully. The main objective of this study was to evaluate the effect of layer structure on physical and mechanical properties of binderless composite plywood. This work also focus on determining the optimum oxidant proportion between veneer and particles

## 2. Material and Methods

Veneer and particles were prepared from Sengon wood (*Paraserianthes falcataria*). The average thickness of the veneer was 0.97 $mm^2$ and particles used were dried particle which passed 10 mesh of sieve screening. Veneer binderless composite plywood was manufactured with oxidation method [8] using 15% of hydrogen peroxide and 7.5% of ferrous sulphate as oxidant. The composite-plywood was produced in size 30 x 30 x 0.7 cm with 0.75 $g/cm^3$ of targeted density. We used two variable productions of veneer layer construction and oxidant proportion. Variation of veneer layer was 3, 5 and 7 layers while oxidant composition applied to veneer and particles layer was 1 : 3, 1 : 4, 1 : 5 and 1 : 6 (w/w). The board were hot pressed for 12 minutes at the temperature of 180°C with pressure of 25 $kgf/cm^2$. Prior testing the specimens, produced composite plywood were conditioned for 2 weeks. Evaluation of physical and mechanical properties of composite plywood were conducted according to Japanese Industrial Standard (JIS) A 5908 2003A 5809-2003[9].

## 3. Result and Discussion

The results showed that the physical properties parameters which fulfilled JIS A 5809-2003 were density and moisture content while all mechanical properties parameters did not fulfill the standard. The physical and mechanical properties of the composite plywood boards which did not meet the required standards were thickness swelling, MoR, MoE, Internal bond and screw holding capacity.

### *3.1 Physical properties*

Figure 1 shows several boards have meet the target density of 0.75g / cm3. This study has been designed to achieve same target density for all the type veneer layer constructions. However the results revealed varying densities. This could be caused by unevenly particles distribution on the mat, causing density variations in some parts of the board. In addition, the uneven distribution of oxidant on the particles caused less bonding between particles and veneer. Another factor which allow variation in the board density was the thickness of sengon veneer which used where the veneer was very thin and undergone shrinking after oven drying. Drying sengon veneer tends to shrink and resulting curved plywood product [10].

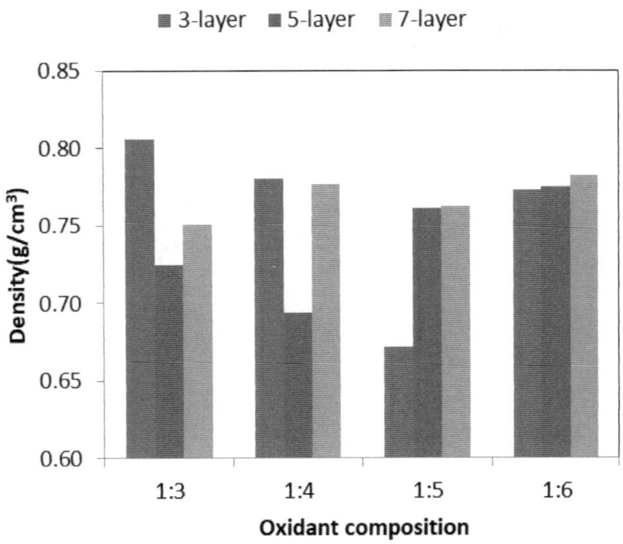


**Figure 1.** Density of the produced composite plywood

Density is one of the important indicators for the quality of composite boards. Density greatly affects most composite board properties [11]. Factor causes density differences is also due to spring back state. The much number of veneer layer used the higher density obtained.
The results of this study revealed that composite plywood with 7 layers has a high average density value. The relation of veneer layer construction on the moisture content value is showed in Figure 2. The more layers are given, the lower moisture content produced. There was clear effect of the layer number on the moisture content.

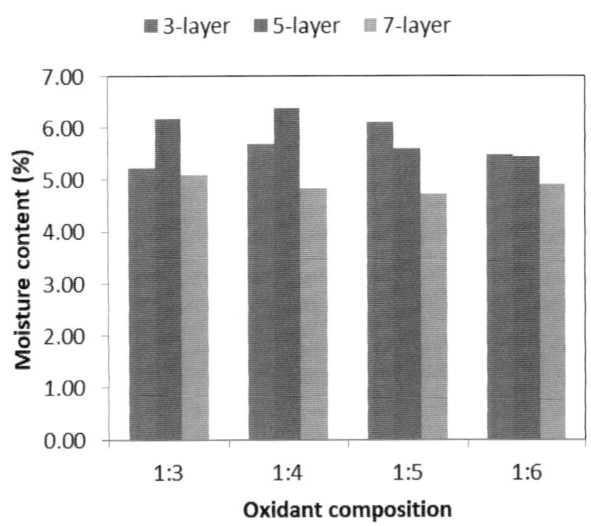


**Figure 2.** Moisture content of the produced composite plywood

Figure 2 shows that the lowest moisture content composite plywood occur on 7 layers of veneers for all type oxidant . proportions. Existing of veneer as a face-back and core decreased moisture content as the result of the less surface of the particles to contact with water due to the covering veneer (face-back). The oxidant composition of 1: 3 and 1: 4 on composite plywood with 5 layers have slightly higher moisture content than that of composite plywood of 3 layers. This could be caused by uneven distribution of the oxidant solution when spread to particles and veneer. However, moisture content value for all the composite plywood construction have met the JIS A 5809-2003 standard, which required of 5-13%. This meant that moisture content did not affect the physical properties of the produced boards.

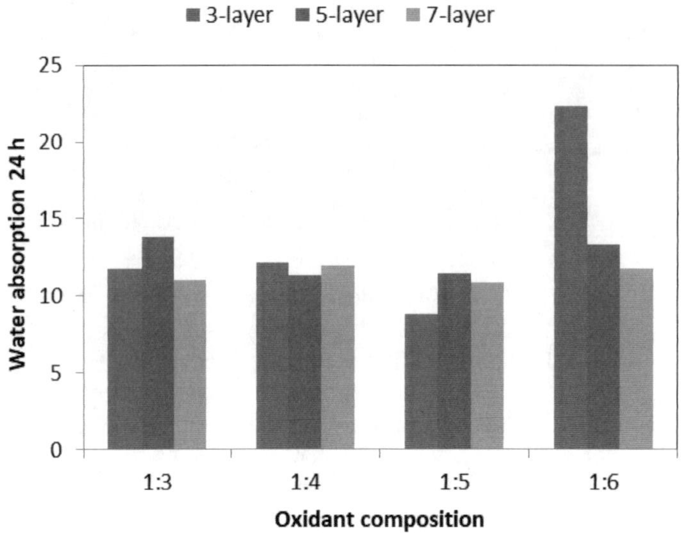


**Figure 3.** Water absorption of the produced composite plywood

Water absorption parameter is not required in JIS A 5809-2003, however, this value related to ability of the board in absorbing water. The value of water absorption composite plywood was founded quite high, ranging 45% -92% for 24-hour water immersion test (Figure 3). The high value of water absorption is influenced by the low density of produced board. One of the most factors affect this value is the number of veneer layers. Increasing number of veneer layer results decreasing water absorption and so in conversely.

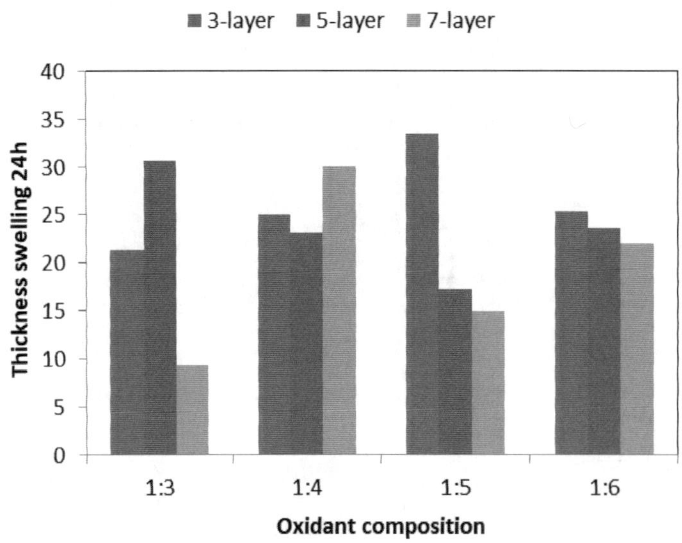


**Figure 4.** Thickness swelling of the produced composite plywood

Thickness swelling is one of the important physical properties because it will determine whether the produced can be used for interior or exterior. The thickness swelling value (Figure 4) of the produced boards did not meet the standard, except of 7-layer board with 1: 3 oxidant composition after 24 hours water immersion test, which requires maximum 12% of thickness swelling. These high value of thickness swelling could be caused by uneven distribution of particles and oxidants which affected weak bonding between particles. Larger thickness swelling and linear expansion values illustrate that produced composite plywood board has low dimensional stability.

*3.2 Mechanical Properties*

Modulus of elasticity values both parallel and perpendicular to grain did not fulfill JIS 5908-2003 which requires values for parallel of 4.08x104 kgf / cm2 and MOE perpendicular of 2.86x104kgf /cm2. It can be seen in Figure 5 and 6, the value of MOE increased with increasing the number of veneer layer. All the composite plywood types with 7 layers and oxidant compositions have high values. Larger MOE value means the board more resistant to shape changes so it requires higher load, otherwise small MOE value means the board does not resistant to shape changes then it requires lower load [12].

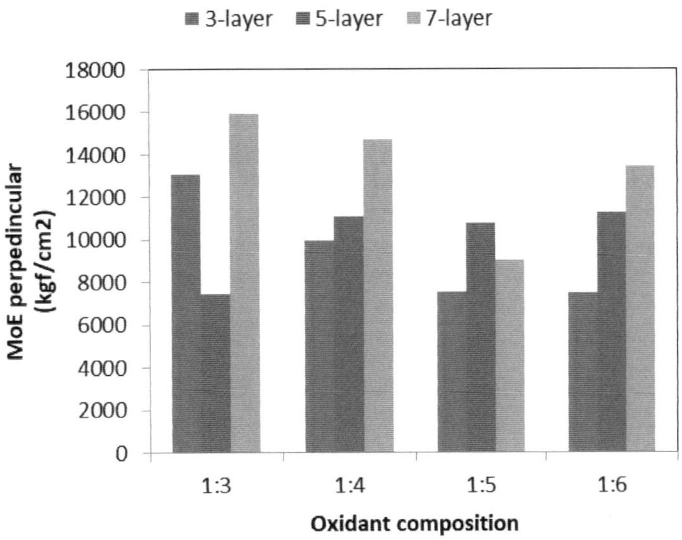


**Figure 5.** MoE perpedincular to grain of the produced composite plywood

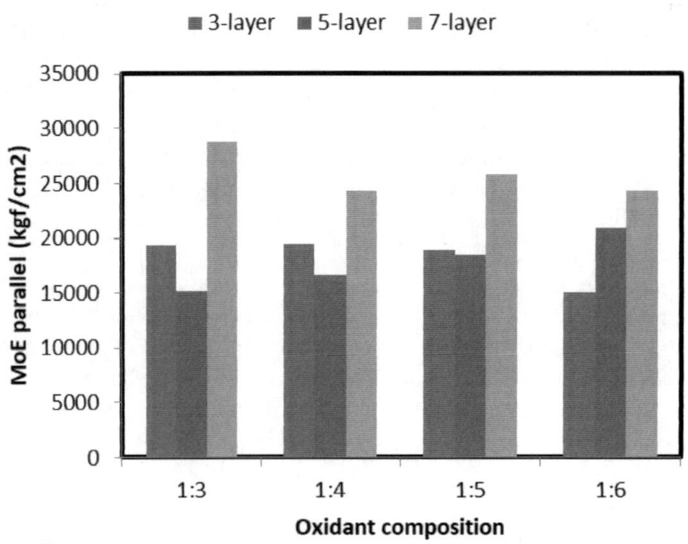


**Figure 6.** MoE parallel to grain of the produced composite plywood

MOR value (Figure 7-8) of produced composite plywood board either parallel or perpendicular to grain, all board type did not meet the standard which required MOR parallel at least 306 kgf/cm2 while MOR perpendicular to grain at least 153 kgf/cm2 . The low value of MOE and MOR is influenced by the veneer raw material, in this case the dimension of sengon veneer was very thin and has curve surface. Uneven oxidant distribution which sprayed throughout particles and veneer was also one of the factors could causes the low values of MOE and MOR.

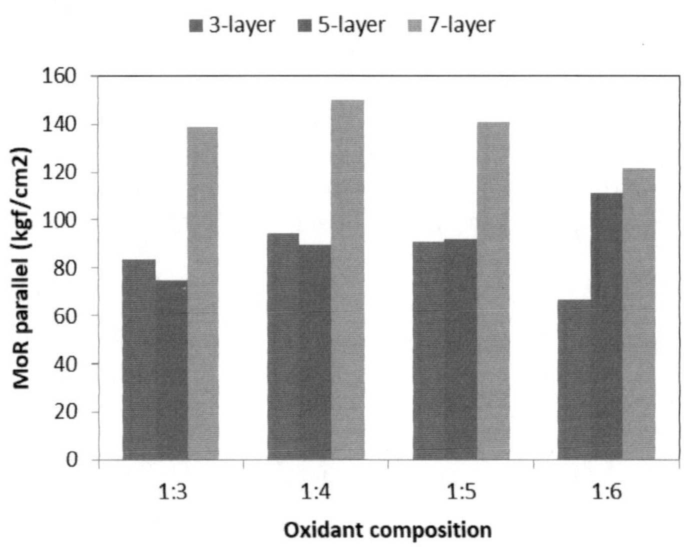


**Figure 7.** MoR parallel to grain of the produced composite plywood

MOE and MOR of produced composite plywood in this study have low values. This could be caused by the present of lathe check on the veneer and would be easily broken when got pressure.

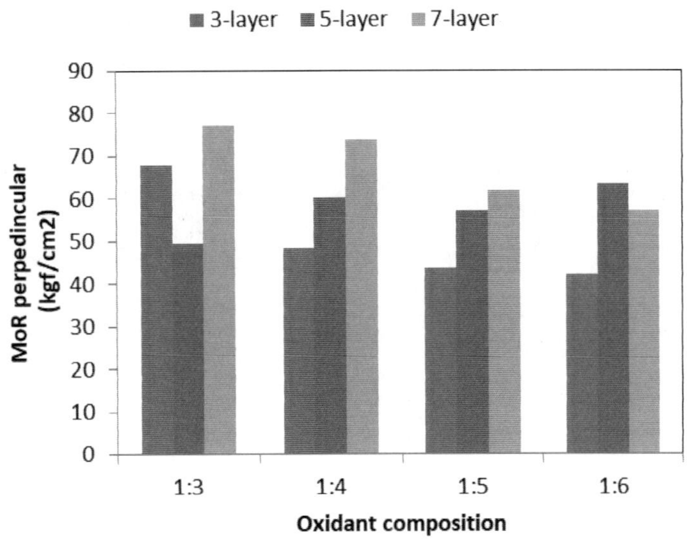


**Figure 8.** MoR perpedincular to grain of the produced composite plywood

Internal bond is one of the most important mechanical properties shows bonds strength between particles and veneers on the board. The internal bond values of the produced composite plywood did not fulfill the standard, which required of 3.1 kgf/cm$^2$ , except of 7-layer board with 1: 3 of oxidant composition. This was similar to other mechanical properties, where increasing number of veneer layers also increase the internal bond of the composite plywood. Figure 9 shows 7-layer board has higher internal bond value than that of 3-layer and 5-layer boards.

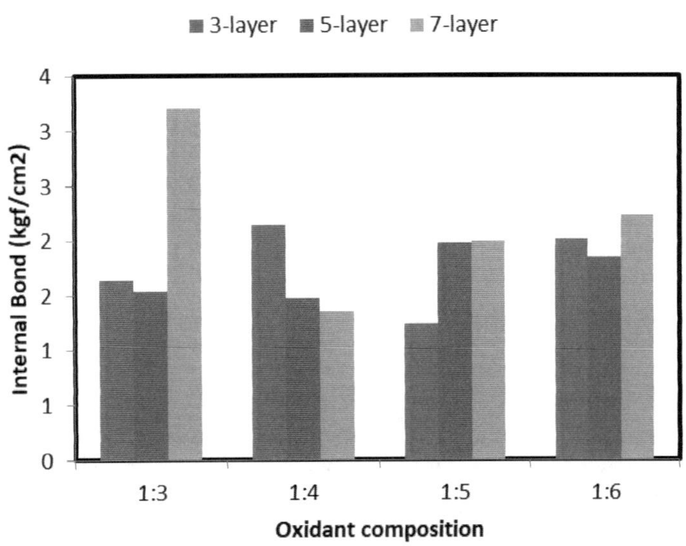


**Figure 9.** Internal bond of the produced composite plywood

The value of screw holding capacity (Figure 10) revealed that all the board types did not meet the JIS A 5809-2003 which required minimum of 0.51 kgf / cm2. The low value of screw holding capacity due to uneven distribution of particles in mat then caused existence of some void and resulted decreasing screw holding capacity. The value of the screw holding capacity varied greatly for each layer and each composition of the oxidant used.

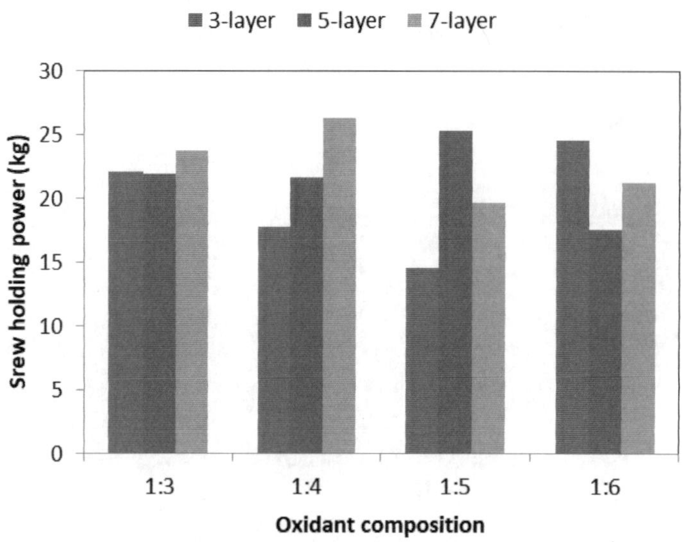


**Figure 10.** Screw holding capacity of the produced composite plywood

The lack characteristics of produced composite plywood in this study could be due to several factors. One of those factors was the thickness of sengon veneer used is very thin with average 0.97 $mm^2$. As for the oxidation process, it was expected that all oxidant composition were able to cover all the particles and veneer, however it occurred in some area in the mat. This caused low adhesion between veneer and particles. The strength of the particle board is basically determined by the strength of the bonds and the strength of each particle [13].

## 4. Conslusions

Physical and mechanical properties of produced composite plywood have fulfilled JIS A 5809-2003 on density and moisture content while thickness swelling, MOR, MOE, Internal bond and screw holding capacity have not met the standard. Layer structure affected composite plywood characteristics and the optimum layer structure was 7-layer while the best oxidant composition between veneer and particles was 1: 3 (w/w). A one weight portion of oxidant was used to oxidize veneer while 3 weight portions were used to oxidize particles.

## References

[1] M H M Amini, R Hashim and N S Sulaiman 2019 Formaldehyde-Free Wood Composite Fabricated Using Oil Palm Starch Modified with Glutardialdehyde as the Binder. *International Journal of Chemical Engineering* 9 pages

[2] Widyorini R, T Higashihara, J Xu, T Watanabe, S Kawai 2005 Self-bonding characteristics of binderless kenaf core composites *Journal of Wood Science* **39** pp 651–662

[3] Widsten P, A Kandelbauer, 2008 Adhesion improvement of lignocellulosic products by enzymatic pre-treatment *Journal of Biotechnology Advances* **26** pp 379–386

[4] Nonaka S, K Umemura, S Kawai 2013 Characterization of bagasse binderless particleboard manufactured in high-temperature range *J Wood Sci* **59** pp 50–56

[5] Saadaoui N, A Rouilly, K Fares, L Rigal 2013 Characterization of date palm lignocellulosic by-products and self-bonded composite materials obtained thereof *Materials and Design* **50** pp 302–308

[6] Nadhari W N A W, R Hashim, S Hiziroglu, O Sulaiman, J G Boon, K M Salleh, M F Awalludin, M Sato, T Sugimoto 2014 Measurement of some properties of binderless composites manufactured from oil palm trunks and Acacia mangium. *J. Measurement* **50** pp 250–254

[7] Suhasman, M Y Massijaya, Y S Hadi, and A Santoso 2011 Particle oxidation time for the manufacture of binderless particleboard. *Wood Res. J.* **2**(1) pp 9–18

[8] Suhasman S, D Yunianti, S Saad and B Baharuddin 2013 Characteristics of Binderless Particleboard Made of Three Species of Sulawesi Bamboos. *Wood Res. J.* **4**(2) pp 68–71

[9] Japanese Standards Association (JSA) 2003 *Particleboards Japanese Industrial Standard (JIS) A 5908-1994* (Japan: Japanese Standards Association)

[10] Kabe A, W Darmawan, M Y Massijaya 2012 Characteristics of Sengon Rotary-Cut Veneer. *J. Ilmu dan Teknologi Kayu Tropis* **10** (2)

[11] Nourbakhsh A 2010 Mechanical and Thickness Swelling of Particleboard Composites Made from Three-Year-Old Poplar Clones *Journal of Reinforced Plastics and Composites* **29**(4) pp 481–489

[12] Makowski A, 2019. Analytical Analysis of Distribution of Bending Stresses in Layers of Plywood with Numerical Verification. *Drvna Industrija* **70**(1) pp 77–8

[13] Maloney T M 1993 Modern Particleboard and Dry-Process Fiberboard Manufacturing Edisi Revisi (USA: Miller Freeman Inc San Francisco)

**Acknowledgement**
The authors acknowledge to the Ministry of Research-Technology and Higher Education of Indonesian Republic, and Hasanuddin University for funding this study.

# Characteristics of the Cement Board Using $CO_2$ Injection Technology from Wood and Non-Wood Species.

**I Taskirawati[1*], A Agussalim[1], B Baharuddin[1], S Suhasman[1]**

[1]Faculty of Forestry, Hasanuddin University, Jl Perintis Kemerdekaan Km. 10 Makassar Indonesia 90245

*E-mail: tasqira@unhas.ac.id

**Abstract**. The fundamental weakness of the technology in manufacturing cement board is that the curing process takes a long time, which is about 28 days. $CO_2$ injection technology has been shown to accelerate the hardening process. However, each of the wood species has a different response due to the presence of inhibitors, namely extractive component. This study aims to compare the characteristics of $CO_2$ injection cement boards from three wood species. These three species came from hardwood, softwood, and non-wood groups, namely agathis (Agathis sp.), Gmelin (Gmelina arborea) and parring bamboo (Gigantochloa atter), respectively. The type of cement used was Portland Cement Composite (PCC). The three cement boards were made using the $CO_2$ injection method. The characteristics of the cement board were tested according to JIS A 5417 – 1992. The results showed that agathis has good characteristics in terms of mechanical properties. Both Internal bonding and modulus of elasticity were better than others, but not with its modulus of rapture. As for physical properties, parring bamboo actually showed good performance. However, in general, these three species of cement boards did not meet the standard.

## 1. Introduction

Cement board is panel products made by combining wood particles or lignocellulose, cement, and water formed by pressing under certain conditions. The raw material used for the manufacture of this product has a wide range in the form of whole wood, as well as other lignocellulose materials such as bamboo, coconut fiber, bagasse, kenaf, hemp, and others. This cement board product can mainly replace plywood and solid wood boards both of which use whole wood raw materials.

Cement boards are potential to be developed as building components that are more economical and more environmentally friendly so that they can be reached by low-income people. Related to this problem, what is interesting to develop is the manufacture of cement boards with $CO_2$ injection technology where this technology can shorten the curing time from previously reaching four weeks to just about one hour. In this process, the injected carbon dioxide will be neutralized quickly by calcium silicate to form carbonate calcium [1], and of course, this makes the cement board denser.

In Indonesia, the development of cement boards is very potential, because in addition to abundant available lignocellulose materials, cement production capacity is also very large, reaching 60 million tons in 2012 and is projected to continue to increase to 75 million tons in 2015 [2]. However, not all lignocellulosic materials can be used as raw material for cement boards due to differences in extractive substances they contain so that the choice of raw materials is limited. In many studies conducted, it was found that in several types of wood containing extractive substances in large proportions, it was able to inhibit the process of setting on cement boards which were characterized by low temperatures of hydration and long hydration times.

Some of the results of research that have been conducted show that the type of raw material used affects the quality of the cement board produced as well. As with conventional cement boards, certain types of cement cannot be used as raw material for cement board construction because the chemical components that make up the soluble wood inhibit the cement hydration process. Therefore, a considerable amount of research has been developed to investigate the compatibility between cement and lignocellulose material.

Research that uses lignocellulosic material in the form of wood and non-wood has also been done. However, no one has compared the use of lignocellulosic material as a raw material for making cement boards using the $CO_2$ injection method. The lignocellulose material used is wood and non-wood particles. The wood particles used are wood originating from community forests. Consisting of gmelina wood to represent hardwood, and agathis to represent softwood. Non-wood lignocellulose material represented by bamboo. This study aims to compare the use of lignocellulosic wood and non-wood materials in the manufacture of cement boards using the $CO_2$ injection method.

## 2. Research Method

### *2.1. Materials*

The lignocellulose material used in this study was agathis (*Agathis sp.*), Gmelina (*Gmelina arborea*) and parring bamboo (*Gigantochloa atter*), obtained from Maros Regency, South Sulawesi Province. This material was ground in mill to obtain particles that pass through a 20 mesh screen and retained on a 40 mesh screen. Cement used as a matrix is a type of Portland Composite Cement (PCC) available on the local market in Makassar while the water used as a solvent is obtained from water managed by PDAM Kota Makassar.

### *2.2. Board manufacturing and $CO_2$ injection*

The cement board was made with a ratio of particles of wood, cement, and water is 1: 2.5: 1.25. Each board was made with a target density of 1.2 g / cm3 and a target thickness of 1 cm. The dough was made by mixing lignocellulose particles (wood or bamboo), cement and water until homogeneous. The mixture was then printed in a mold measuring 25 cm x 25 cm x 1 cm on an iron plate coated with transparent plastic. The mixture that has been prepared in the mold on the iron plate was then pressed to a thickness of 1 cm. The iron plate containing the pressed mixture is isolated around it and left at room temperature for 24 hours to reach the setting process.

After 24 hours, the board that has undergone the setting process was then given a $CO_2$ injection treatment. Wood is injected $CO_2$ for 60 minutes to reach curing. $CO_2$ injection is done by first inserting the board into the injection tube. In the tube, $CO_2$ flows. The condition was then maintained for 60 minutes. After that, the board is removed from the tube and put in the desiccator for ± 15 minutes. The board was then weighed and conditioned at 80°C 10 hours.

### *2.3. Testing*

The boards produced from the conventional method and $CO_2$ injection are then tested to be tested for physical and mechanical properties concerning the Japanese Industrial Standard (JIS) A 5417-1992 [3]. Besides JIS A 5417 - 1992, the standard was used according to Bison's patent [4]. Testing physical properties include; density, moisture content, water absorption, thickness swelling, and linear expansion while testing mechanical properties include modulus of rupture (MoR), modulus of elasticity (MoE), and internal bond

## 3. Results and Discussion

Comparison of the characteristics of the cement board where the lignocellulosic material comes from softwood (Agathis), hardwood (Gmelina) and non-wood (Bamboo parring). The three boards are made using the $CO_2$ injection method. The characteristics of the cement board tested in this study were the physical and mechanical properties of the cement board. Comparison of these characteristics can be seen as follows:

### *3.1. Physical Characteristics of Cement Board*

#### *3.1.1. Water content*

The moisture content of particle wood and non-wood cement board is shown in Figure 1. The non-wood particle cement board has a higher moisture content (3.65%) than the wood particle cement board. Research conducted by Bakri and Suhasman [5] also showed that Bamboo pariing is a type of bamboo that has the highest water content compared to other species of bamboo. The cement board of hardwood (Gmelina) has a lower moisture content (0.67%) compared to softwood cement board (Agathis) of 1.87%. Suhasman [6] in his research compared Gmelina wood with other wood species. The water content of Gmelina wood obtained is 7.18%, but the method used is different because this study uses a conventional method with the addition of $CaCl_2$. While the research was conducted by 7.2%. However, the three cement boards still have water content below 16%. Based on JIS A 5417-1992 which requires the board's water content to be a maximum of 16%, the three types of cement board made meet these standards.

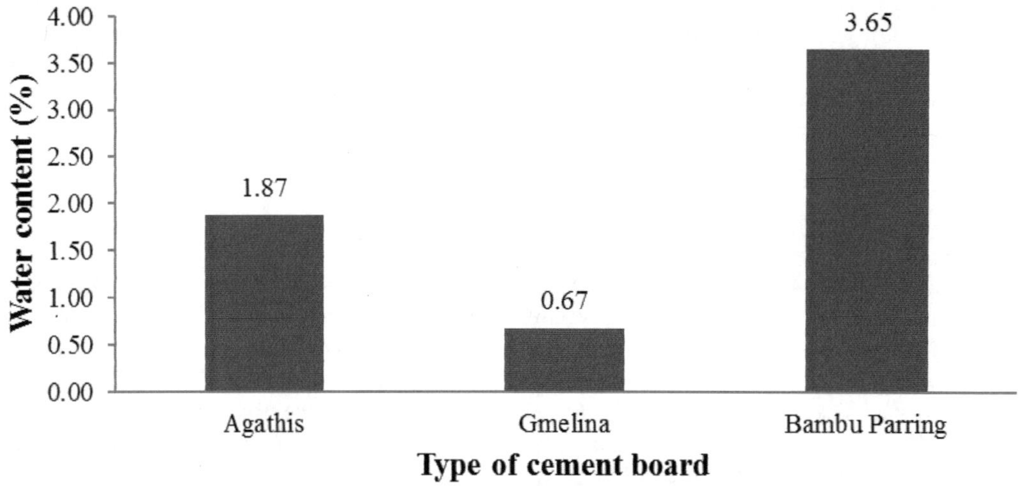


**Figure 1.** Water content of cement board particle wood (Agathis and Gmelina) and non-wood (Bamboo parring).

#### *3.1.2. Density*

The density values of wood and non-wood particle cement boards are shown in Figure 2. The non-wood and wide hardwood particle cement boards have a higher density, each 0.95 g/cm$^2$, and 0.94 g/cm$^2$, compared to softwood particle cement boards. However, none of the three cement board samples reached the desired density value of 1.2 g/cm$^3$. It is because the thickness of the three boards tested exceeds the set target of 1 cm, while the average thickness of the board is 1.26 cm. As a result of the volume of the board also becomes larger, so that the three boards have a low density and far below the desired target. Bakri and Suhasman [5] stated that another possibility is that the board thickness exceeds the desired target is the setting of the hydration process only for 24 hours, while there is still water left on the cement board. This water is still used for the hydration process to get the maximum amount of calcium silicate (maximum amount of calcium silicate hydrate) for 28 days at room temperature.

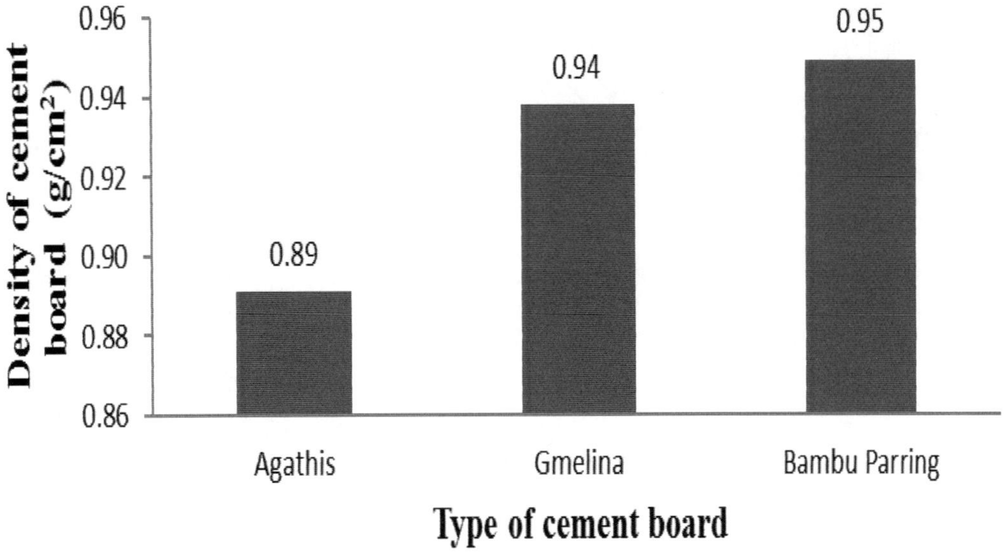


**Figure 2.** The density of cement board particle wood (agathis and gmelina) and non-wood (bamboo parring).

*3.1.3. Water absorption*

The results of the measurement of water absorption on wood and non-wood particle cement boards are shown in Figure 3. The value of water absorption on cement boards soaked for 2 hours ranged from 49.96 - 47.51%. The value of water absorption on cement boards soaked for 24 hours ranged from 54.65 - 50.20%.

The value of water absorption of non-wood particle cement board (from 2 times, 2 hours and 24 hours) obtained lower than the water absorption of wood particle cement board. Meanwhile, the value of the water absorption of softwood particles is higher than the water absorption capacity of cement boards of hardwood particles. According to Muharam [7] the closer the contact between particles, the more water vapor will be harder to get into the particle board. So that the higher the density, the more absorption the water will be. It is in line with the results of the density value in Figure 2, where the density of non-wood particle cement boards is high so that the water absorption is low.

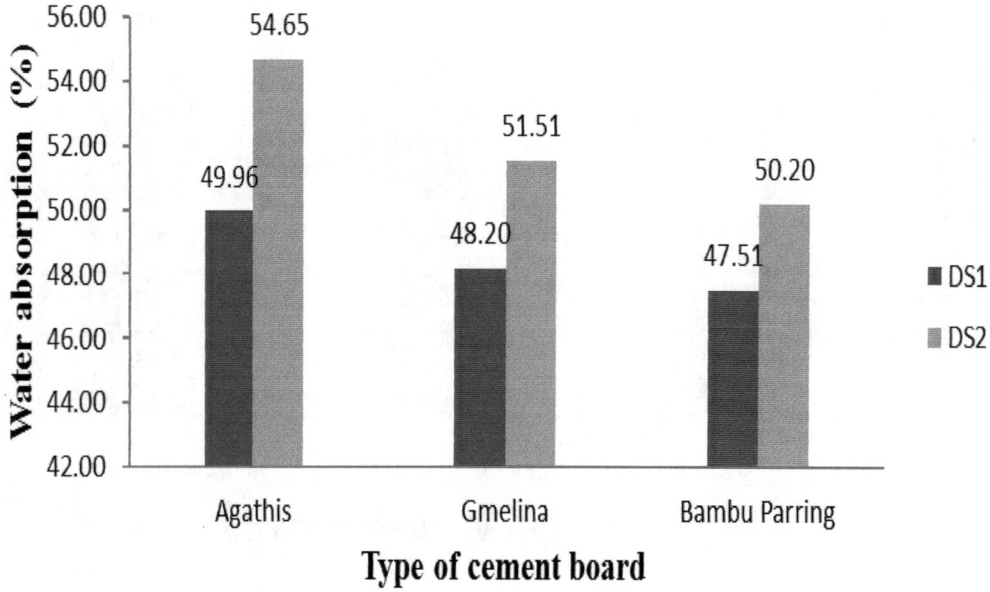


**Figure 3.** The water absorption of cement board particle wood (Agathis and Gmelina) and non-wood (Bamboo parring).

*3.1.4. Thickness swelling and linear expansion*

The value of thickness swelling and linear expansion cement boards soaked in water for 2 hours and 24 hours is shown in Figure 4. The thickness swelling and linear expansion cement boards of hardwood particles soaked for 2 hours are higher than the cement board of softwood and non-wood particles. The value of thickness swelling and linear expansion softwood particles cement soaked for 24 is lower than that of hardwood and non-wood particle cement boards.

Riyadi [8] suspects that there is a relationship between water absorption and thickness swelling. It is because the more water is absorbed and enters the fiber structure, the more significant the change in dimensions produced. However, the thickness swelling and linear expansion values obtained on wood and non-wood particle cement boards in this study are inversely proportional to water absorption.

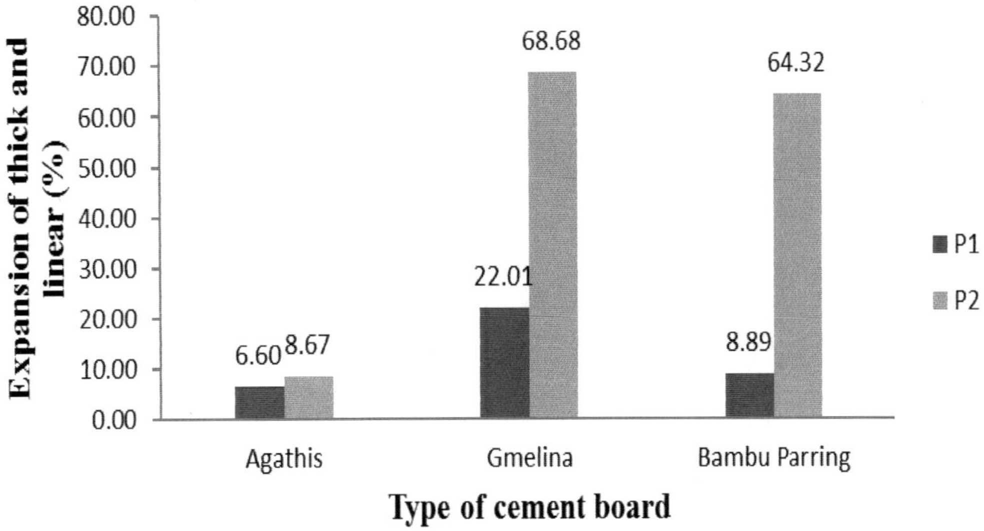


**Figure 4.** The thickness swelling and linear expansion of cement board particle wood (Agathis and Gmelina) and non-wood (Bamboo parring).

### *3.2. Mechanical properties of cement board*

#### *3.2.1. Internal bond*

The internal bond values of wood and non-wood particle cement boards are shown in Figure 5. The internal bond values of cement hardwood and non-wood wood particles are lower (0.6 kgf /cm$^2$ and 0.8 kgf/cm$^2$ respectively) compared to softwood particle cement boards (1.5 kgf/cm$^2$). Based on JIS A 5417-1992 which requires a minimum adhesive strength of 3.10 kgf/cm$^2$, there is none of the sticky firmness (IB) cement board for wood and non-wood particles that meet this standard. The same thing happened in the study conducted by Bakri and Suhasman [5] and Suhasman [6] where each of them examined bamboo cement boards (0.41 kgf/cm$^2$) and gmelina cement board (1.34 kgf/cm$^2$).

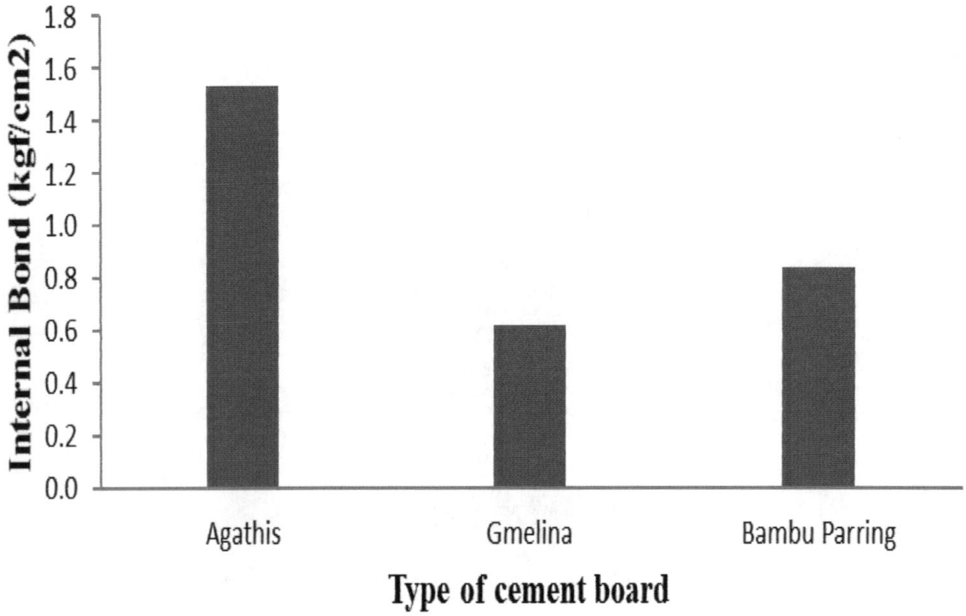


**Figure 5.** The Internal Bond of cement board particle wood (Agathis and Gmelina) and non-wood (Bamboo parring).

*3.2.2. Modulus of rupture (MoR)*

The modulus of rupture (MoR) cement board of wood and non-wood particle are shown in Figure 6. The calculation results obtained from the MoR cement board wood particle is higher (16 kgf/cm$^2$ and 19 kgf/cm$^2$) when compared to MoR cement board of non-wood particle. Based on JIS A 5417-1992 which requires a minimum MoR value of 63 kgf/cm$^2$, the MOR value of wood and non-wood particle cement boards does not meet these standards. It is caused by the board density value that does not meet the desired target of 1.2 g/cm$^3$. and 80.37 kgf/cm$^2$.

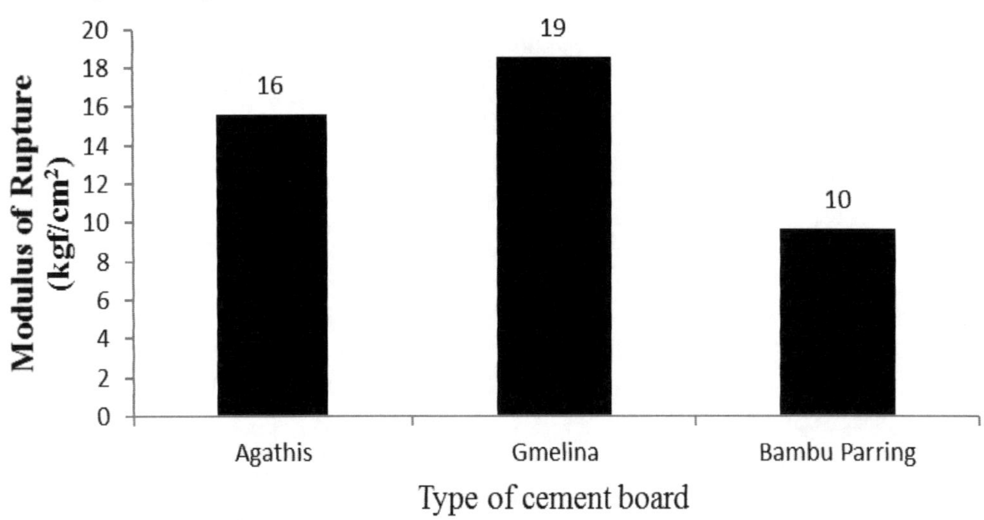


**Figure 6.** The modulus of rupture (MoR) of cement board particle wood (Agathis and Gmelina) and non-wood (Bamboo parring).

*3.2.3. Modulus of Elasticity (MoE)*

The MoE values of wood and non-wood particle cement boards are shown in Figure 7. The calculation results showed that MoE value of cement board from softwood is higher (3835 kgf/cm$^2$) compared to MoE of cement board from hardwood and non-wood wood particles (each 2292 kgf/cm$^2$ and 2083 kgf/cm$^2$). Research using the same material (bamboo parring and gmelina) was also carried out by Bakri and Suhasman [5] and Suhasman [6] with results of 2338.82 kgf/cm$^2$ and 13784 kgf/cm$^2$ respectively.

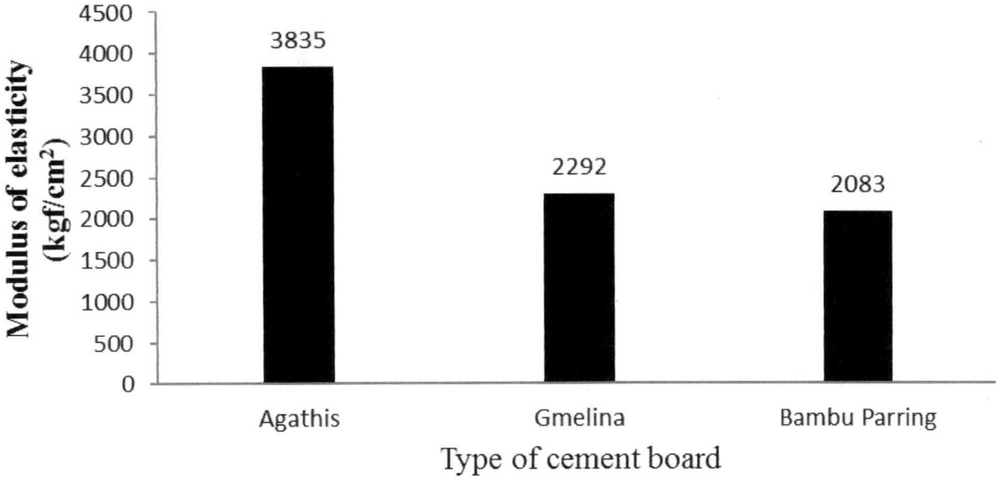


**Figure 7.** The modulus of elasticity (MoE) of cement board particle wood (Agathis and Gmelina) and non-wood (Bamboo parring).

The MoE value of the three cement boards does not meet Bison's commercial standards. Where the standard requires the MoE value of the cement board to be more than 30000kgf / cm2 [4].

## 4. Conclusion

- Agathis has good characteristics in terms of mechanical properties.
- Both Internal bonding and modulus of elasticity were better than others, but not with its modulus of rapture.
- For physical properties, parring bamboo actually showed good performance. However, in general, these three species of cement boards did not meet the standard.

## References

[1] Na B, Wang Z, Wang H, Lu X. 2014 Wood-Cement Compability Review. *Wood Research* **59** (5) pp 813–826
[2] Kementerian Pekerjaan Umum, 2013 *Statistik Semen Nasioanal* (Indonesia: Statistik Semen Nasional)
[3] [JSA] Japanese Standards Association. 1992 *Cement Bonded Particle Board* (Japan: Japanese Industrial Standard (JIS) A 5417-1992)
[4] Bison 1975 Cement-Bonded Particleboard Plant Integrated With Low Cost Housing Production Unit Case Study Prepared for FAO Portofolio of Scale Forest Industries for Develoving Countries. Germany: Bison Werke Bahre and Breten Bmtt and Co. 3257 Spring IFR.
[5] Bakri, dan Suhasman 2014 Carbon Dioxide Injection in Bamboo Cement Board Manufacturing. *Jurnal Ilmu dan Teknologi Kayu Tropis* **12**(1) pp 82–90

[6] Suhasman 2012 Comparison of Characteristics of Cement Boards from Stems and Branches of Wood from Community Forests – in Bahasa. *Jurnal Perennial* **8**(1) pp 30–35
[7] Muharam A 1995 Effect of Particle Size and Density on Sheet Physical Properties and Mechanics of Sugarcane Waste Particle Board – in Bahasa. (Bogor: unpublish).
[8] Riyadi C 2004 Physical and mechanical properties of fiber boards from banana stem waste (*Musa sp.*) At various initial treatments and paraffin content – in Bahasa. (Bogor: unpublish).

**Acknowledgment**
This research is funded by the Ministry of Research Technology and Higher Education of the Republic of Indonesia through the scheme of Superior Research of Applied Higher Education in fiscal year 2018 We would like to thank the University Hasanuddin and LPPM Hasanuddin University for the support given to this research, and Heru Arisandi, Deni Edriani Junaedi, and Nurfianah Mustamin for their assistance during the research.

IOP Conf. Series: Materials Science and Engineering **593** (2019) 012014 doi:10.1088/1757-899X/593/1/012014

# Analysis of Absorption Properties of a Composite FlyAsh and $Fe_2O_3$ for X-ray Radiation Shielding Applications

**M A Budiawan[1], S Suryani [2*], B Abdullah[3] , D Tahir[4]**

Department of physics, Hasanuddin University, Makassar, 90245 Indonesia

*Email: suryani@fmipa.unhas.ac.id

**Abstract**. X-ray radiation shielding materials are generally made of lead. The inconvenience of using lead is heavy and expensive. Therefore, research has been conducted on X-ray radiation shielding materials made from a mixture of fly ash and $Fe_2O_3$, which is lighter, and cheaper, because the material is easily obtained. The results showed that radiation shielding material made from a mixture of fly ash and $Fe_2O_3$ had good X-ray radiation energy absorption for material thickness of 2 cm and for low X-ray radiation energy.

## 1. Introduction

X-ray is one of the radiation which has the ability to ionize like gamma and ultraviolet radiation. X-rays have a frequency region of $3 \times 10^{16}$ Hz to $3 \times 10^{19}$ Hz, so X-rays have energy from 100 eV to 100 KeV. The use of X-rays in the health sector is a necessity, therefore X-ray instruments can be found in hospitals to private clinics. The extent of the use of X-ray instruments starting from conventional X-ray instruments, CT scan which is used for examination of the limbs, and X-rays which is used for dental examinations. That is why, it certainly requires good protection to keep the exposure dose of X-ray radiation to public remains low.

Various kinds of materials are used to minimize X-rays exposure, and lead is the best radiation shielding material, because lead which has density of 11.34 g / $cm^3$ can absorb all X-rays energy with a power of 100 keV only with a thickness of 2 mm [1]. Other materials that can also absorb X-rays radiation with the same ability as lead with a thickness of 2 mm are concrete with a thickness of 15 cm and a brick with a thickness of 25 cm [2], combined of sodium pentaborate and barium sulfate [3]. Another X-rays radiation shield used as an apron is made from a cotton cloth coated with silicone rubber consisting of a mixture of tungsten, bismuth or barium sulfate powder with the same heavy fraction [4]. Even in 2016 a study was conducted by Seon Chil Kim who combined barium sulfate with silicone fluids to form an apron, and the result showed that the apron had the ability to absorb the same energy with lead, but the weight of the apron was only one-fifth of lead apron [5[. Actually, X-rays radiation shields on the market today, usually made from a mixture of lead that is quite heavy, or glass fiber which is quite expensive, so it is necessary to obtain a cheap and easily obtained shielding material.

Material that is easily obtained, because this material is a waste that can pollute the environment is fly ash. Fly ash is residual combustion of coal, which contains silica or silica + alumina compounds. In a very fine form, fly ash can react with calcium hydroxide (at sufficient humidity and room temperature) to form a material that has binding or solid.The binding properties of the fly ash are then used as additives in concrete mixtures and cement making mixtures. Fly ash generally has the form of of solid, round, or round hollow particles, or round particles that contain other smaller round particles. solid, round, or round hollow particles, or round particles that contain other smaller round particles. This material has a size that varies from less than 1 μm to more than 100 μm (some literature mentions a size of 0.5 μm - 300 μm), but mostly we find less than 20 μm [6]. Researchs about utilization of fly ash as a radiation shield material have been done by researchers from all over the world, for example a research was carried out by Harjinder Singh Mann et.al in 2018, who made clay fly ash bricks as gamma-ray shielding materials [7], and Kanwaldeep et.al tries to compare the use of a mixture of cement in concrete with a mixture of fly ash in concrete as a shielding materials for gamma-ray radiation [8].

Other material which is easily obtained in nature is iron. Iron does not exist in the form of atoms, but in the form of molecules consisting of oxygen and iron atoms which are bonded together, known as iron ore. Iron ore itself is usually obtained in the form of magnetite ($Fe_3O_4$), hematite ($Fe_2O_3$), goethite, limonite or siderite. Iron can be used as a mixture to form a radiation shield, while iron itself is not good to be used as a shield of X-rays radiation, because radiation shielding materials must have a large atomic number, such as lead.

The utilization of fly ash as a radiation shield without mixture is not good, because the form of fly ash which is round will form a cavity in its solid form. Therefore, a mixture is needed, so that the shielding material can be dense and not easily broken. In this study, a mixture of fly ash with ($Fe_2O_3$) was used, with the aim to obtain X-rays radiation shielding material that is cheap, easily found, not easily destroyed, and preserves the environment through the use of coal burning waste.

## 2. Material and Method

Fly Ash (FA) used in this study is fly ash with type C or has CaO concentration more than 20%. It is taken from the Jeneponto Steam Power Plant - Jeneponto Regency and Iron ore or $Fe_2O_3$ which has a grain size of 5-9 μm with a purity of 99.5%.

20 grams of fly ash is mixed with 0.20 grams of $Fe_2O_3$. The mixture is binded with an alkaline solution which is a mixture of $Na_2O_3.SiO_2$, NaOH, and $H_2O$. After the mixture has thickened and can be formed, the mixture is poured in containers of size (6 cm x 6 cm x 2 cm), (6 cm x 6 cm x 4 cm), and (6 cm x 6 cm x 6 cm), and heated at 70°C for 1 hour and let it rest for 24 hours at 28°C.

Radiation exposure to samples is carried out using Siemens X-rays instrument with the type of multimobile digital as a source of ionizing radiation. The detector used is Pen Dosimeter. The X-rays plane tube is set at a fixed voltage of 90 Kv, current of 32 and 40 mA, and the distance between focus and material is 100 cm, the Pen Dosimeter is placed under the object, the radiation area has size area of 25 cm x 25 cm. The exposure for each material which has a different thickness is then carried out by putting the center of the beam right at the center.

When light or radiation passes through the medium, we can observe a decrease in the intensity of light or transmitted radiation. This is because some of the intensity of light or radiation is absorbed by the medium. Radiation absorption by the medium follows the Lambert-Beer law, which can be written as:

$$I = I_0 \, e^{-\mu x} \tag{1}$$

with $I_0$ is the initial intensity, I is the intensity transmitted, x is the thickness of the medium, and μ is the coefficient of absorption or attenuation.

The ability of reducing X-rays energy by radiation shielding material is presented by half value layer (HVL) parameter. The HVL parameter can be determined using the equation which can be written as :

$$\mathrm{HVL} = \frac{0.693}{\mu} \tag{2}$$

### 2.1 Result and Discussion

The sample is a mixture of fly ash with $Fe_2O_3$, with the same area of 36 $cm^2$, but different thicknesses are 2, 4 and 6 cm (Figure 1). Furthermore, the measurement of density was carried out, and we obtained 2.21 g / $cm^3$.

1 2 3

**Figure 1.** Samples thickness of (1) 6 cm, (2) 4 cm, and (3) 2 cm.

After sample is dry, the sample is then irradiated with X-rays. X-rays instrument is arranged to operate at a maximum electrical voltage of 90 kV and two electric current values of 32 mA and 40 mA. The rate of radiation dose coming out of the X-rays instrument was measured with the Pen Dosimeter detector was 930 µGy/s for a 32 mA electric current, and 931 µGy/s for a 40 mA electric current.

**Table 1.** Sample characteristics at irradiation with currents of 32 mA.

| Samples | Thickness (cm) | Voltage (kV) | Current (mA) | Rate of radiation dose (µGy/s) | Rate of radiation dose transmitted (µGy/s) | Linear attenuation ($cm^{-1}$) | HVL (cm) |
|---|---|---|---|---|---|---|---|
| 1 | 2 | | | | 330 | 0.518046 | 1.34 |
| 2 | 4 | 90 | 32 | 930 | 135 | 0.482477 | 1.44 |
| 3 | 6 | | | | 85 | 0.398756 | 1.74 |

Result did not show a significant difference in the rate of radiation dose in the setting of a low electric current with a higher one. The size of linear attenuation is calculated using the equation in the theory and linear attenuation values showed that there was an influence of medium thickness on attenuation values, where increasing thickness of the material decreases attenuation values or reduces the ability of the medium to absorb X-rays radiation.

**Table 2.** Sample characteristics at irradiation with currents of 40 mA.

| Samples | Thickness (cm) | Voltage (kV) | Current (mA) | Rate of radiation dose (µGy/s) | Rate of radiation dose transmitted (µGy/s) | Linear attenuation ($cm^{-1}$) | HVL (cm) |
|---|---|---|---|---|---|---|---|
| 1 | 2 | | | | 412 | 0.407618 | 1.70 |
| 2 | 4 | 90 | 40 | 931 | 170 | 0.425115 | 1.63 |
| 3 | 6 | | | | 105 | 0.363716 | 1.91 |

We also observed that the increase in X-rays energy which was in this research presented by the increase in electric current could reduce the ability of the medium to absorb X-rays radiation or material attenuation (Table 1and 2) (Figure 2). This is due to the composition of the material can affect the interaction between the material and the photon, thus affecting the attenuation of the material. This was also observed by Danial Salehi who found that the ability of soft tissue attenuation depends on the energy of the photons [9].

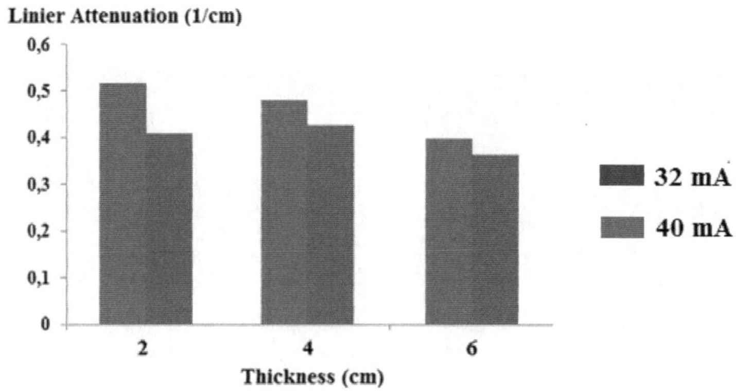


**Figure 2.** Material attenuation as a function of material thickness.

The ability of the material to absorb radiation energy is shown by the half value layer (HVL) parameter. The highest HVL is seen in high material thickness and for high X-rays radiation energy, whereas the lowest HVL value is found in materials with low material thickness and for low X-rays radiation energy. As a result the radiation shielding material that can absorb X-rays radiation energy is a material with a thickness of 2 cm for low radiation energy.

## *3.* Conclusions

The best radiation shielding materials for X-rays radiation are materials with large atomic numbers, such as lead, but lead has a large density, so radiation shields which are made of lead will be heavy, and furthermore the price of lead is also expensive. To overcome this, research on the use of materials that are easily obtained in the environment is carried out and can also preserve the environment. The material is fly ash which is a waste and $Fe_2O_3$ which is easily obtained. The results showed that radiation shielding material which is a mixture of fly ash and $Fe_2O_3$ that can absorb X-rays radiation energy is a material with 2 cm thick for low radiation energy

## References

[1] McCaffrey J.P., H. Shen, B. Downton and E. Mainegra-Hing 2007 *Medical Physics*, **34** (2) p 530
[2] Susilo, I N Rahma, Mosik and Masturi 2017 *J. Phys.: Conf. Ser.* **824** 012005
[3] Tahsin Elmas 2016 *International Journal of Advances in Science Engineering and Technology* **4** (2) p 22
[4] Nebahat Aral, F. Banu Nergis and Cevza Candan 2016 *Textile Researh Journal* **86** (8) p 803
[5] Seon Chil Kim, Jeong Ryeol Choi and Byeong Kyou Jeon 2016 *Scientific Reports* **6,** 27721
[6] Khairul Nizar Ismail, Kamarudin Hussin and Mohd Sobri Idris 2007 *Journal of Nuclear and Related Technology* **4** p 47
[7] Harjinder Singh Mann, Gurdarshan Singh Brar, Kulwinder Singh Mann and Gurmel Singh Mudahar 2018 *Nuclear Engineering and Technology* **xxx** 1
[8] Kanwaldeep Singh,Sukhpal Singh and Gurmel Singh 2014 *Journal of Energy* 1
[9] Danial Salehi, Dariush Sarden and M.S. Jozani 2015 *Journal of Radiation Research and Applied Sciences* **8** (3) p 439

# Characteristic of Chitosan Adhesive from Shell Shrimp Litopenaeus vannamei and Their Application For Producing Particleboard

**D W Ningsi[1*], Suhasman[1], S Saad[1]**

[1]Faculty of Forestry, Hasanuddin University, Makassar, Indonesia

*Email: dwi.wahyuningsih33@gmail.com

**Abstract.**The shrimp shell that rich compounds of chitin can be converted to produce polysaccharides called chitosan. The chitosan can be utilized as bioadhesive for substitute chemical based adhesive for producing particleboard. The purpose of this study was to evaluate the effect of different concentration acetic acid ($CH_3COOH$) as the solvent to made chitosan adhesive and physical and mechanical properties of the produced particleboard by using sawdust waste. Chitosan adhesive formulation will be used as an adhesive by dissolving chitosan powder with $CH_3COOH$ solvent. This research focused on optimization the solvent content wich is $CH_3COOH$ in different concentration such as 0.5%, 2% and 4%. Particleboard with size of 25 cm x 25 cm x 0.7 cm with 8%, target density 0.75 g/cm$^3$ were produced using sawdust waste of Paraserianthes falcataria. The processing condition were 180°C in temperature during 12 minutes with pressure of 25 kg/cm$^2$. The research showed, there is no particular trend in particle board characteristics related to solvent concentration. Physical and mechanical Properties fulfill requirement of JIS A 5908-2003 for Moisture Content, MOE and IB. The produced particleboard that using chitosan as bioadhesive and sawdust waste as particle is potentially to be developed technically based on the quality of the particleboards produced.

## 1. Introduction

South Sulawesi is one of shrimp production resource which able exported in large number. The volume of shrimp were exported from 11 shrimp industry in South Sulawesi at 2016 reached out 6,884,5 tonnes. The exported shrimp are the frozen formed shrimps without skin and the potential waste of shrimp skin left was quite large. Therefore, the use of shrimp skin residues is very potential to be used as a raw material for chitosan adhesives.

Research on chitosan has been done by some researchers such as investigate raw material, temperature and reaction time on characteristics of chitosan produced [1]. Sinaga (2009) study the effect of chitosan by modifying the stages of demineralization, deproteination, and deacetylation. Adhesive application using chitosan with maple wood particles [2] and utilizing chitosan as an adhesive for producing particleboard from Sengon wood [3]. A number of those researchers using acetic acid ($CH_3COOH$) to dissolve the chitosan to used as an adhesive. Based on research for the use of chitosan, the concentration of solvents applicated at chitosan has never been studied before. Therefore this study focuses to answer the problems related to it.

Other main materials needed for the production of particle board are wood particles or other lignocellulosic materials. The advantage, the particle board does not require raw materials of high quality particles. For example, coconut fiber [4], rattan wastes [5] and the collection waste of acacia wood [10] as a material for the production particle board.

In other side, plenty wood waste from sawmill industry remains less use. Waste from sawmill industries often left to pile up, decomposed and burned. The average volume of sawn timber in 2014-2015 in South Sulawesi is 87785 m3 [6]. The amount of waste produced by the sawmill industry was on average 40.48% per year [14]. The high volume of waste produced can cause environmental pollution if not controlled. So, this study focuses on the effect of different concentration of $CH_3COOH$ solvents to chitosan in the manufacture of particle boards by utilizing sawn wood waste.

## 2. Materials and Method

The wood particles used were waste particles from the sawmill and then were milled to particle by using knife ring flaker. The proportion of sawdust particles size was identified by 6 screens levels: passed 6 mesh/hold 9 mesh, passed 9 mesh/hold 12 mesh, passed 12 mesh/hold 22 mesh, passed 22 mesh/hold 40 mesh, passed 40 mesh/hold 60 mesh, and passed 60 mesh/hold 80 mesh, respectively. The proportion of the particles size distribution data shown in Figure 1. Furthermore, the particles were dried until the moisture content was about 4%.

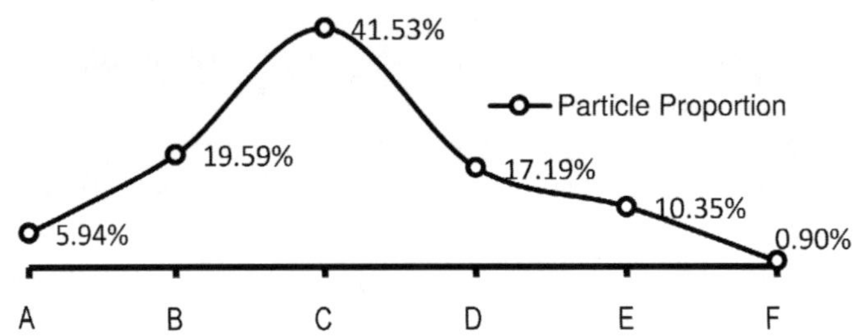


**Figure 1.** Particles size: (A) Passed 6/hold 9 mesh, (B) Passed 6/hold 12 mesh, (C) Passed 12/hold 22 mesh, (D) Passed 22/hold 40 mesh, (E) Passed 40/hold 60 mesh, (F) Passed 60/hold 80 mesh

The shell shrimp (*Litopenaeus vannamei*) were obtained from PT. Multi Monodon Indonesia. The shell shrimp then washed from residual meat were still attached and air dried and crushed with *hammermil*. There are 3 (three) steps for producing chitosan, namely Demineralization, Deproteinization, and Deacetylation [8].

### *2.1 Chitosan Production*

Demineralization of shrimp shell has been carried out with concentration HCl 2 M with a solid solvent ratio 1:15 (w/v) into a baker. The solution was heated at 70°C for 4 hours while stirring every 10 minutes. The residual was washed with water flows until neutral pH then the shell shrimp were air dried. Deproteinization of shrimp shell was done with 3.5% NaOH with a solid solvent ratio 1:10 (w/v) at 70°C for 4 hours while stirring every 10 minutes. Protein Solution was removed and washed with water flows until netral pH. Then purified chitin was air dried.

The chitosan was prepared by removal of acetyl groups from chitin. It was experimented using concentration of NaOH 60% with solid solvent ratio 1:20 (w/v) at 110°C for 4 hours while stirring every 10 minutes. The residual was washed until neutral pH with water flows. The resulting of chitosan then dried and prepared for characterization.

### *2.2 Characterization of chitosan produced*

Quality of chitosan produced was checked following by Protan Laboratory Standard consist of rendemen, moisture content, viscosity, solubility, solution colour, and deacetylation degree.

#### *2.2.1 Rendemen*

Rendemen of chitosan was determined according to the following equation:

$$\text{Rendemen (\%)} = \frac{\text{Output}}{\text{Input}} x\ 100 \tag{1}$$

where output is the chitosan weight and input is the chitin weight before heated for deacetylation

*2.2.2 Moisture content*

Moisture content of chitosan was determined by *moisture analyzer.* About 2 g chitosan was put inside *mositure analyzer*. Then the measured values showed was calculated according to the following equation:

$$\text{Moisture content (\%)}= \frac{w0-w1}{w1}\ x\ 100 \quad (2)$$

where w0 is wet weight of the sample and w1 is dry weightof the sample.

*2.2.3 Ash content*

Ash content measured with about 2 g chitosan put into pourcelain then it was dried at 100°C in the oven to constant weight. The sample was placed in a desicator and measured the weight. Furthermore, the sample was put on a muffle furnace and heated at 750°C for 6 hours. Then placed in a desicator for minimum of one hour. The sample was allowed to cool and the weight of ash residue was weighed. The ash content was calculated according to the following equation:

$$\text{Moisture content (\%)}= \frac{w1}{w0}\ x\ 100 \quad (3)$$

where w0 is dry weight sample after ovened and w1 is weight of ash residue.

*2.2.4 Solubility*

Solubility of chitosan was checked with dissolved 2 g chitosan in 2% Acetic Acid with solid solvent ratio 1:100 (w/v). Then the sample was filtered using vacuum filtration and chitosan residue was heated at 100°C for 1-2 hours until get a constant weight. The sample placed in a desicator and the weight was weighed. Solubility of chitosan was calculated according to the following equation:

$$\text{Solubility (\%)} = \frac{w0-w1}{w0}\ x\ 100 \quad (4)$$

where w0 is initial weight of chitosan before dissolved and w1 is final weight of chitosan insoluble after heated and get a constant weight.

*2.2.5 Viscosity*

Viscosity of chitosan and chitosan adhesive was determinated using *Brookfield Viscometer.* Chitosan solution was prepared in 1% acetic acid at 1% concentration on dry basis. The value are reported in centipoise (cP). For chitosan adhesive, 30 g chitosan dissolved in different concentration of acetic acid (0.5%, 2%, and 4%) with solid solvent ratio 1:5 (w/v). Then the value are reported in centipoise (cP).

*2.2.6 Degree of deacetylation*

Degree of deacetylation (DD) determined with base line method using FTIR (*Fourier Transform Infrared Spectroscopy*) instrument with frequency range of 4000–400 $cm^{-1}$. The DD of chitosan sample was calculated based on the following equation:

$$\text{DD} = \left(1 - \left(\frac{A1655}{A3450} X \frac{1}{1.33}\right)\right) X\ 100\% \quad (5)$$

where A was log (P0/P) for absorbance, A1655 and A3450 were the absorbance at 1655 cm−1 of the amide-I band as a measure of the N-acetyl group content and 3450 cm−1 of the hydroxyl band as an internal standard to correct for thickness.

*2.2.7 Particleboards production*

Chitosan adhesive was made using dissolved chitosan powder with different concentration of acetic acid (0.5%, 2%, and 4%) with solid solvent ratio 1:5 (w/v). The solid content was used at 8% based on the weight of the oven-dried particles. Furthermore, chitosan gel was distributed into the particles until evenly mixed. The mixture was hand-formed into a mat by using a forming box of 25 x 25 cm with a target density of 0.75 g/cm2. The metal plate was put in the side of the mat to control the board thickness to 0.7 cm. Then the mat was given preliminary pressure and the forming box was removed. After that, the mat that has been formed was pressed using a hot press, which was sealed with thin stainless steel frame. The pressure, temperature and time of press were 25 kg/cm2, 180°C, and 12 min, respectively. Three replications of each production were performed in this study. Then particleboard were conditioned at room temperature for about 14 days.

*2.2.8 Evaluation of board properties*

The conditioned particleboard tested were physical (moisture content (MC), density (D), thickness swelling (TS), and water absorption (WA)) and mechanical properties (modulus of elasticity (MOE), modulus of rupture (MOR), and internal bond (IB)) according to Japanise Industrial Standard for particleboard (JIS A 5908-2003). The MC and D test were performed on a 10 x 10 cm specimen. The TS and WA test were performed on a 5 x 5 cm specimen from each board after they underwent water immersion for 24 h at room temperature. The weight and thickness of the specimens were recorded for each specimen before and after immersion. The IB was test in spacimens of the same size of those used. The bending properties of the boards were evaluated by a bending test on a 15 x 5 cm specimen for each board in dry conditions.

## 3. Result and Discussion

The properties of chitosan can be observed qualitatively and quantitatively. According to the Standard Protan Laboratory there were several criteria to evaluate the quality of chitosan [4, 19]. The properties of chitosan according to the Standard Protan Laboratory and the results of the study showed in Table 1.

**Tabel 1.** Properties of chitosan according to the Standard Protan Laboratory and research results.

| Parameters | Standard Protan Laboratory | Chitosan produced | Remark. |
|---|---|---|---|
| The yield (%)* | | | |
| a. Shell shrimp powder-demineralization | | 29.61 | |
| b. Demineralization-deproteination (chitin) | - | 84 | - |
| c. Deproteination-deacetylation (chitosan) | | 82 | |
| d. Shell shrimp powder-chitosan | | 20.6 | |
| Moisture content (%) | ≤ 10 | 4.8 | Fulfilled |
| Ash content (%) | ≤ 2 | 0.55 | Fulfilled |
| Viscosity (cP) | | | |
| a. Low | <200 | | |
| b. Medium | 200-799 | 44.32 | Low |
| c. High | 800-2000 | | |
| d. Very High | >2000 | | |
| Solubility in 2% acetic acid | Soluble | Soluble | Fulfilled |
| Deacetylation (%) | ≥ 70 | 64.02 | Unfulfilled |
| Viscosity of chitosan adhesive (cP)* | | | |
| a. Solvent consentration 0.5% | | 4.752 | |
| b. Solvent consentration 2% | - | (E)** | - |
| c. Solvent consentration 4% | | (E)** | |

*) Not specified in the Standard Protan Laboratory

*3.1 The yield*

The chitosan production process started from demineralization process, which is removing the minerals content at the shell shrimp using HCl solution. In this stage, the minerals contained will reacted with HCl. The effervescence were appear during of heating the shell shrimp was indicated as a reaction of HCl with minerals salt. In this stage, the yield obtained was 29.61%. Furthemore, result of demineralization process was reacted with NaOH solution on waterbath therefore protein contents were dissolved in NaOH solution [9]. A little of effervescence were formed on the surface of the solution, condenced and reddish colour. The condenced was occured because the protein contents were released and linked with Na+ ions and resulted natrium proteinat. The ions Na+ were binding the end of the negatively charged protein chain until settled [9]. After this stage the remaining content of the shrimp

skin powder was chitin. The yield of chitin in this work was 84%. Isolation of chitosan compounds was obtained by performing a deacetylation reaction on chitin. The chitosan was extracted by converting the acetyl groups into amine group in chitin using strong alkalis (NaOH). Therefore, the yield of chitosan in in this study was 82%. The chitosan yield from shell shrimp was 67.08% which is higher than their study [8].
Therefore, the yield of chitosan obtained from this study begined on the shell shrimp powder until chitosan was 20.60%. The calculated of chitosan was not required on the Standard Protan Laboratory but useful for calculating the amount of chitosan that will be applied.

### *3.2 Moisture and ash content*

The moisture content in the present study according to the past works [3, 8]. The commercial chitosan products required by less than 10% moisture content. It was well within the range of chitosan produced. Generally, the moisture content contained in chitosan expressed as $H_2O$ which is bound to polymer functional groups especially amine groups, N-acetyl and hydroxyl through hydrogen bonds [9]. The moisture content of chitosan can be influenced by the process during drying, the drying time, the amount of dried chitosan and the surface in which the chitosan was dried [8].

The ash content of chitosan was indicates as inorganic compounds contained in the raw material of chitosan. The ash content of chitosan in this study and commercial chitosan by Standard Protan Laboratory were 0.55% and less or same than 2%, respectively.

### *3.3 Viscosity and solubility*

The value viscosity of chitosan in this study was 44.32 cP. The value was in low category according to Standard Protan Laboratory. Information about viscosity of chitosan was related to its application. In the pharmaceutical field was required chitosan with low viscosity, whereas for the purposes of thickening or hardening necessary groceries chitosan with high viscosity [11].

The solubility of chitosan on acetic acid is also one of the parameters in determining the quality of chitosan. The higher solubility of chitosan, the better quality of the chitosan produced [8]. In this study, solubility of chitosan was 94.37% which was fulfilled in Standard Protan Laboratory required. Several factors affected the solubility of chitosan were temperature, concentration of alkaline (NaOH), ratio of chitin and alkaline (NaOH) and the size of chitosan powder [12].

### *3.4 Deacetylation degree*

The deacetylation process was carried out to transform chitin into chitosan. Deacetylation was the process to removed acetyl group in chitin to produced chitosan. The amount of acetyl group missing from chitin was called the degree of deacetylation (DD). DD of chitosan can be indicated from function group formed used FTIR analysis. The results of functional group analysis and FTIR reference are shown in Tabel 2.

| **Funcional Group** | **Wave number ($cm^{-1}$)** | |
|---|---|---|
| | **FTIR reference * ($cm^{-1}$)** | **Chitosan produced ($cm^{-1}$)** |
| OH | 3448 | 3444.87 |
| C-H | 2891.1 | 2881.65 |
| $NH_2$ scissoring, N-H bending | 1655.0 | 1653 |
| $NH_2$ wagging and twisting | 850.0-750.0 | 896 |

*)[12; 1]

Chitosan FTIR spectra showed the emergence of OH strech which indicated absorption in the wave number area 3444.87 cm-1 and amide (C = O) at the 1660.71 cm-1 wave. disappearance of C=O groups in waves 1680-1660 cm-1 indicates the loss or reduction of the acetyl group on chitosan [13]. The reaction of chitosan formation from chitin is a hydrolysis reaction of an amide from a base. Chitin acts as an amide and NaOH as a base. In this process, the -OH group enters into the NHCOCH3 group and then eliminates the $CH COO^-$ group and resulting in an amine, namely chitosan [14].

The chitosan DD obtained in this study was 64.02%, The result did not satisfy the Standard Protan Laboratory. According to Pujiastuti (2001), the DD chitin compared to chitosan usually ranges from

70-100% depending on its use. The low chitosan of DD from the results of this study may be caused by the non-optimal isolation process, especially in the stirring process. The stirring factor and heating temperature used were influence the chitosan DD [8].

*3.5 Viscosity of adhesive*
In this study chitosan was used as an adhesive for production particleboard. The viscosity value of the chitosan formulated in tadhesives is presented below:

**Tabel 3.** Viscosity of chitosan adhesive

| Solvent concentration of $CH_3COOH$ (%) | Viscosity (cP) |
|---|---|
| 0.5 | 4.752 |
| 2 | E* |
| 4 | E* |

*)*Over range*

Based on the results test of adhesives with a solvent concentration of 0.5%, the viscosity value was 4752 cP. While for solvent concentrations 2 and 4%, the viscosity value were illegibled because the maximum viscosity that can be measured is 12,000 c,so the value were over ranged. It is happened because dissolved chitosan in concentrations 2 and 4% resulted too condenced adhesive. The viscosity of chitosan adhesive was higher with increased the solvent concentration [15].

*3.6 Characteristic of particleboard*

*3.6.1 Physical properties*
Determination physical properties of particleboard in this study were moisture content, density, thickness swelling and water absorption.

*3.6.1.1 Moisture content*
The average value of moisture content of the particle board in this was range about 7.9–8.5%. The relationship between chitosan solvent concentration and moisture content of the particleboard showed in Figure 3.

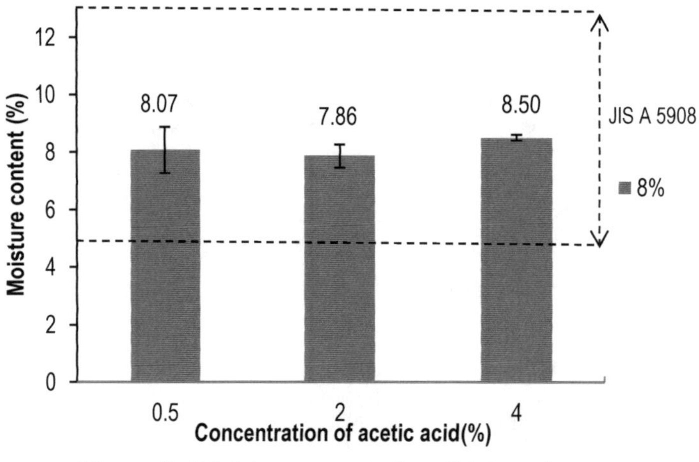


**Figure 2.** Moisture content of particleboard

Figure 2 showed that the moisture content value of particleboard in this study were the treatment with a solvent concentration 2% and 4% for the lowest and the highest value, respectively. The moisture content of the particleboard in this study was satisfied with JIS A 5908-2003 that required the value in ranged 5–13%. The moisture content of the particle board depends on the surrounding air conditions. In addition, wood particles are hygroscopic the adhesive used (chitosan) also has hydrophilic properties [9].

*3.6.1.2 Density*

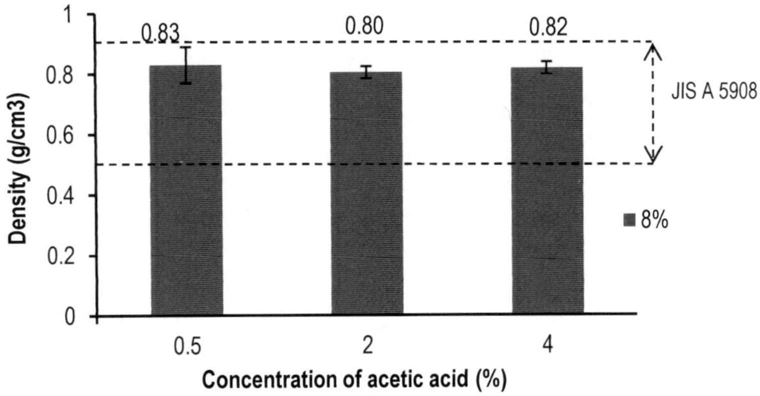


**Figure 3.** Density of particleboard

Based on figure 3 it can be seen that the average density value produced in this study ranged from 0.80 to 0.83 g/cm$^3$. The lowest density was obtained on particleboard using a solvent concentration of 2% and the highest density was found in a treatment with a solvent concentration of 0.5%. JIS A 5908-2003 requires the densities of a good particleboard produced between 0.5-0.9 g /cm$^3$. Overall, the density of the particleboard made in this study met the standard, but not in fulfilled with the targeted density.

*3.6.1.3 Thickness swelling*

The results of the particle panel test showed that the average of thickness swelling after being immersed for 24 hours at around 17.6-25.8%. The highest and the lowest value thickness swelling was found on the board using 2% and 0.5% solvent concentrations, respectively. The value of the thickness swelling obtained can be seen in Figure 4.

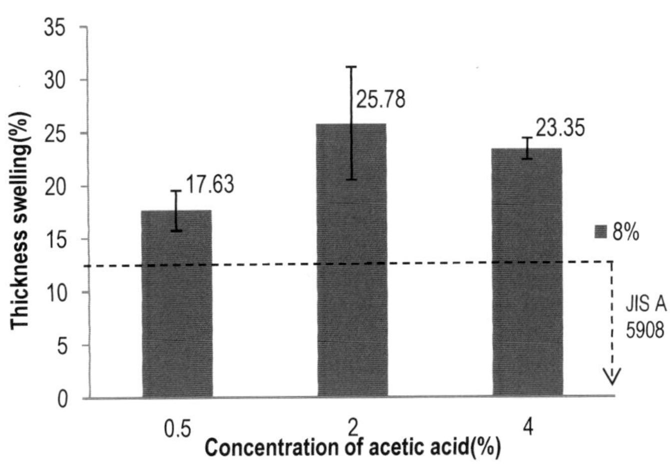


**Figure 4.** Thickness swelling of particleboard

Based on JIS A 5908-2003, the resulting particleboard does not fully met the standards that require 12%. The absence of the certain trends in thickness swelling that associated with of solvents concentration may be caused the adhesive was not dispersed and the moisture content of the particles and the excess of adhesive. Visual observations showed there was different color in the center of the particleboard indicated the particles and adhesive water was trapped during the pressing process.

*3.6.1.4 Water absorption*

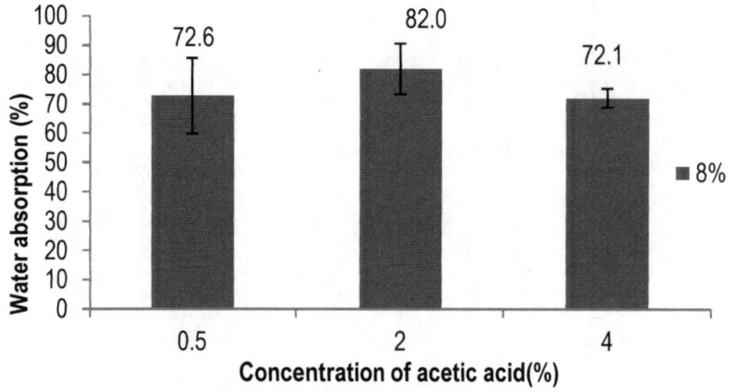


**Figure 5.** Water absorption of particleboard

Water absorption is the physical properties table that shows the board ability to absorb water. The average water absorption value obtained in this study ranged from 72.1 to 82%. The highest water absorption was produced by the particleboard using 2% solvent concentration and the lowest water absorption on the particleboard with a solvent concentration of 4%. The value of water absorption is not required in JIS A 5908-2003. Decreased water absorption indicated, the adhesive enters into empty spaces inter- particles. Therefore, the space that can be inserted by the water was decreases [16].

### *3.6.2 Mechanical properties*

#### *3.6.2.1 Internal bond (IB)*

IB shows the bond value between the particles so the internal bond can be used as a good reference to determine the quality of the particleboard produced [17]. The average IB value of the particleboard ranges from 3.4 to 5.5 kg/cm$^2$. The highest value was found on the board using a solvent concentration of 0.5%. The IB value can be seen in Figure 7. The highest value was produced by the board using a solvent concentration of 0.5% while while the board using solvent concentrations of 2% and 4% almost resulted the same value.

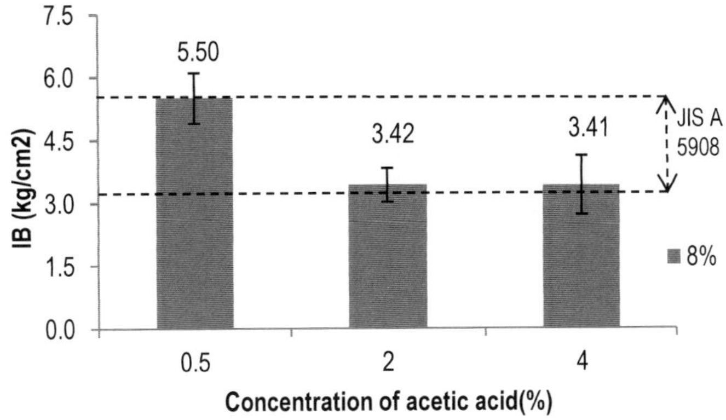


**Figure 6.** Internal bond of particleboard

Data in Figure 6 showed the IB value of the particleboard tends to decrease with increase in solvent concentration. This probably caused of the high viscosity of the adhesive with solvent concentrations of 2% and 4% (Table 3). Viscosity greatly affects the quality of an adhesive. The higher the adhesive viscosity, the ability of the adhesive to flow, move, retain penetration and wetting were lower, its caused the quality of the bonded product was also low [18]. JIS A 5908-2003 required the internal bond of the particleboard is at least 1.5 kg/cm$^2$. Therefore, the IB value of the resulting particleboard has met the standard.

#### *3.6.2.2 Modulus of elasticity (MOE)*

The young's strength test is performed to show the strength of the particleboard to withstand the compressive forces. However, this parameter important because the utilities of particleboard in furniture always requires in flat usage [9]. The MOE value of the particleboard is shown in Figure 7. The average MOE values resulted ranged from 22983 to 24472 kg/cm$^2$. The highest MOE value was produced by the board using a solvent concentration of 0.5% and the lowest value on the board using a solvent concentration of 4%.

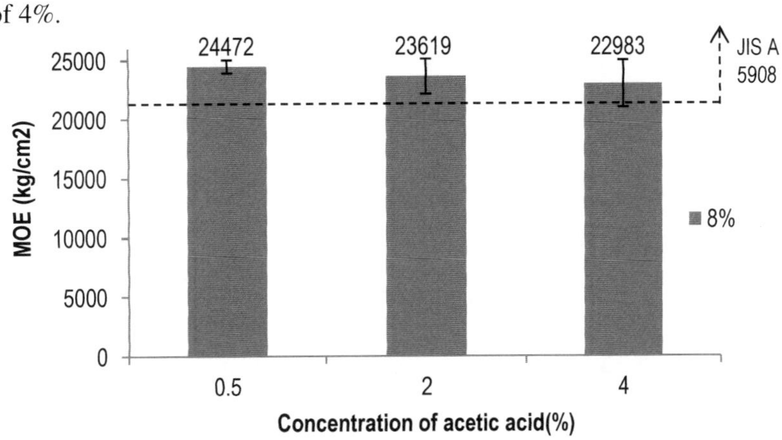


**Figure 7.** Modulus of rupture of particleboard

Data in Figure 8 showed that the highest concentration of solvent used, then resulted MOE value lower. Based on JIS A 5908-2003 which required a minimum bending value at least 20400 kg / cm$^2$, so all the particleboard produced met the standard.

*3.6.2.3 Keteguhan Patah (MOR)*

The Rupture streght is a measure of the maximum load that can be received by the wood. The average values of the MOR ranged from 95.82 to 166.72 kg/cm$^2$. The MOR value is shown in Figure 8.

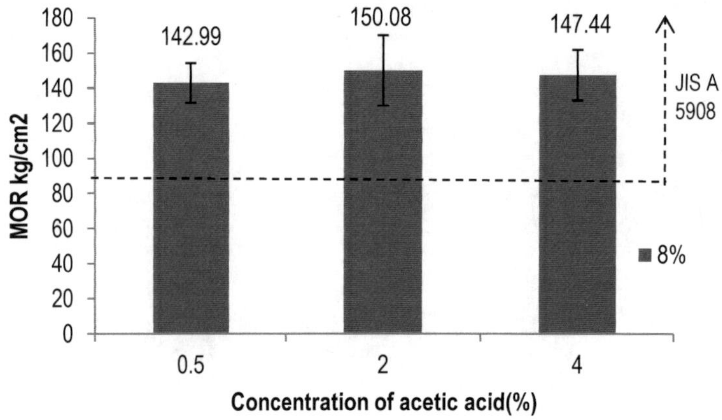


**Figure 8.** Modulus of rupture of particleboard

The highest MOR value was resulted on particleboard using 2% solvent concentration while the lowest MOR value was on the particle oard using a solvent concentration of 0.5%. Compared to the JIS A 5908-2003, as a whole the chitosan-adhesive particleboard has met the standard which requires a minimum MOR value of 82 kg/cm$^2$.

*3.7. General discussion*

The results of this study did not show a particular tendency on the particle board characteristics related to solvent concentrations. The solvent used in this study was $CH_3COOH$ solution. The use of this solution is caused insoluble of chitosan in water. Therefore it is necessary to add $CH_3COOH$ in water. Theoretically, the concentration of the solvent will affect the solubility of adhesive, thus ultimately increasing the adhesive dispersion by increasing the concentration solution.

However, this phenomenon was not found in this study. Variations in solvent concentrations associated with dimension change, adhesive strength and tensile strength, the best values were found on particleboard using a solvent concentration of 2%. While the best value for the elasticity is on the panel with a solvent concentration of 0.5%. It is caused the DD value produced in this study is lower than the required standard, at least 70% so the expected quality objective can be achieved with a lower solvent concentration. The value of DD must be higher than 60% to be soluble in acid even for about 50% of the solvent used is neutral ph [18].

Overall, even the resulting value is not the best, but considering the technical, standard, and economic aspects, the concentration of the solvent at 0.5% is the optimal treatment.

## 4. Conclusions

The characteristics of the chitosan from the shrimp skin tested in this study met almost all the standards specified in the Protan Standard Laboratory except for DD%. The use of chitosan adhesive in particleboard production shows that the optimum concentration of the solvent is adhesive using a concentration of 0.5% $CH^3COOH$. Except for the thickness swelling parameters, all the physical and mechanical properties of the particle board parameters fulfill the standards set by JIS A 2908 - 2003.

Based on the data of the tested parameters, it is indicated that the production of adhesives from shell shrimp is very feasible and has the potential to be technically developed based on the quality of the particleboard produced.

**Reference**

[1] Srijanto B, Imam P, Masduki Purwatiningsih 2006 Effect of Degrees of Deacetylation of Raw Materials in Chitosan Depolymerization in Indonesian *Media-BPPT Akta Kimindo* **1**(2) pp 67–72

[2] Basturk M 2012 Heat Applied Chitosan Treatment on Hardwood Chips to Improve Physical and Mechanical Properties of Particleboard. Departement of Forest Product Engineering, Faculty of Forestry, University of Kahramanmaras Sutcu Imam, Turkey. *BioResources* **7**(4) pp 4858–4866

[3] Kartini 2016 Effect of NaOH concentration on the degree of deacetylation of chitosan as a bioadhesive and characteristic of the particleboard produced in Indonesian. Makassar Faculty of Forestry Hasanuddin University

[4] Sudarsono, Rusianto T, Yogi S 2010 Manufacture of Coconut Fiber Raw Particle Boards with Natural Binder (Kopal Glue) in Indonesian. Electrical Engineering, Faculty of Technology Industry. *Journal of Technology* **3**: 22–32

[5] Suroto 2010 Effect of Size and Concentration of Adhesives on the Physical and Mechanical Properties of Rattan Waste Particle Boards in Indonesian. Baristand Industry of Banjarbaru. *Journal of Forest Product Industry Research* **2** pp 18–30

[6] Badan Pusat Statistik 2016 Wood forest production by type production in South Sulawesi Province ($m^3$), 2013-2015 in Indonesian.https://sulsel.bps.go.id/dynamictable/2016/08/10/177/produksi-kayu-hutan-menurut-jenis-produksi-di-provinsi-sulawesi-selatan-m3-2013-2015.html. Accses on 19 December 2018

[7] Purwanto D 2009 Analysis of Types of Wood Waste in Wood Processing Industry in South Kalimantan in Indonesian. Baristand Industri Banjarbaru, South Kalimantan. *Journal of Forest Product Industry Research* **1**(1) pp 14–20

[8] Agustina S, I Made D, I Nyoman S 2015 Isolation of chitin, characterization and Sythesis Chitosan from shell shrimp in Indonesian. Medical study, Udayana University. *Journal Chem.* **9**(2) pp 271–278

[9] Dompeipen E J, Marni K, Riardi P M 2016 Isolation Of Chitin And Chitosan From Waste Of Skin Shrimp E-journal.kemenperin.go.id/bpbiam

[10] Widiyanto A 2011 Quality of Rubber Wood Particle board (*Hevea brasiliensis Muell.* Arg) and Bambu Tali (*Gigantochloa apus* Kurz) with Wood Liquida Adhesive in Indonesian. Research Institute for Agroforestry. *Journal of Forest Product Research* **29** (4) pp 301–3011

[11] Saputro A N C, Indriana K, Sutarno 2009 Effect of Isolation Method on the Characterization of Chitosan (Surakarta: Faculty of Marine Science and Fisheries, National University of Surakarta)

[12] Hossain M, Iqbal A. 2014. Production and characterization of chitosan from shrimp waste. Bangladesh. Bangladesh Agricurtural University **12**(1) pp 153–160

[13] Puspawati N M, I N Simpen 2010 Optimization of deacetylation of chitin from crab skin being Chitosan from restaurant waste through variation of NaOH concentration in Indonesian. Udayana University. *Journal of Chemical* **4**(1) pp 79–90

[14] Mahatmanti W 2001 Adsorption Study of Zinc (II) and Lead (II) Metal Ions on Chitosan and Chitosan-Sulphate from Windu Shrimp Shells in Indonesian (Yogyakarta: Universitas Gadjah Mada)

[15] Mati-Baouche, Pierre-Henri E, Helene D B, Guillaume P, Cedric D, Philippe M 2014 Chitosan as an adhesive *France European Polymer Journal*

[16] Mawardi I 2009 Quality of Particle Boards of Palm Oil (KKS) Based on Polystyrene Adhesives. State Polytechnic Mechanical Engineering Department National Polytechnic of Lhokesemawe, Banda Aceh in Indonesian. *Journal of Mechanical Engineering*.**11**(2) pp 91–96

[17] Suherti Farah D NurhaidaPhysical and mechanical properties of Durian (*Durio Sp*) skin particles with different concentrations of formaldehyde urea in Indonesia. (Kalimantan: Faculty of Forestry. Tanjungpura University)

[18] Rinaudo M 2006 Chitin and Chitosan: Properties and applications. Prancis: Universitas Joseph Fourier. *Journal of Polymer Science* **31** pp 603–632

IOP Conf. Series: Materials Science and Engineering **593** (2019) 012016 doi:10.1088/1757-899X/593/1/012016

# Mathematical Modelling Time Dependent for Heat Transfer Process in Homogenous Composite ZnO/Activated CNF

**M F Latief[1], D Tahir[1*], E Juarlin[1], Suarga[2]**

[1]Department of Physics, Hasanuddin University, Makassar 90245 Indonesia
[2]College of Computer Science of Dipanegara, Makassar 90245 Indonesia

*Email: dtahir@fmipa.unhas.ac.id

**Abstract.** A theoretical study of heat transfer on composite Zinc Oxide (ZnO)/Activated Carbon Nanofibers (CNF) by using mathematical model from diffusion equation for stationary composite with various heating time was studied. The equation and boundary conditions are interpreted by using Taylor series expansion to obtain a steady-state numerical solution from the Jacobi iteration algorithm. The composites where put inside the furnace model by filaments on the two sides of the walls as the source of heat (general model for furnace). By using this model, the position of sample to receive the homogeneous temperature determined. We compared the steady-state and unsteady-state conditions to identify the temperature distribution at real-time. The results show that the heat distribution will reach an asymptotic state by performing a numerical analysis is 200 K and 300 K for 25 minute time of heating. In this study, shows the efficient way for annealing homogenous materials to identify the position inside the furnace which can be used as a guide to receive the homogenous temperature during the annealing processes.

## 1. Introduction

Treatment of materials during and or post growth by annealing processes is used to identify the temperature stability of the materials which is the key factor in device applications and performance. Annealing processes at high temperatures usually used in industrial production and also for research in laboratory scale. Recently, there are many scientific reported that the composite of ZnO/CNF (carbon nanofibers) used annealing treatment for production of carbon nanofibre as well as in composite form which are candidate materials for supercapacitor electrodes [1]. Carbon nanofibers (CNF) are carbon material with interconnected three-dimensional mesoporous texture and very good thermal stability which was promising electrode material for supercapacitors [2-10]. The temperature of annealing is the important parameter was strongly influenced to the properties such as surface area, pore size and distributions, which are play an important properties in the performance of electrochemical capacitance [8-21]. The sample position inside the furnace is the key factor in receiving the homogenous temperature of the sample in laboratory scale as well as also in the industrial production.

Therefore, in this paper, a theoretical approach was used to describe the temperature distribution of composites ZnO/activated CNF by using diffusion equations. The assumption of the composite ZnO /CNF is homogeneous particles and the distribution is uniform. The purpose in this study is to establish a mathematical model which is suitable for the distribution temperature in the furnaces by determining the optimal conditions of the system. The effects of continuous annealing are producing the heat conduction distribution in the system and also the convection flows from material to air due to the system is un-isolated from the environment. The heat transfer processes during the annealing in the furnace model will be analyzed by numerical system.

## 2. Mathematical Model for ZnO/CNF Composite Heating Process

### *2.1 Model and Assumption*

In this paper, we describe the annealing model of homogenous material in the furnace with filaments as heating sources continuously propagate conduction to the composite ZnO/CNF. We assumed that in the heating processes there no gap between the filament and the material, so there is no difference in thermal conductivity distribution in the ZnO/CNF composite.

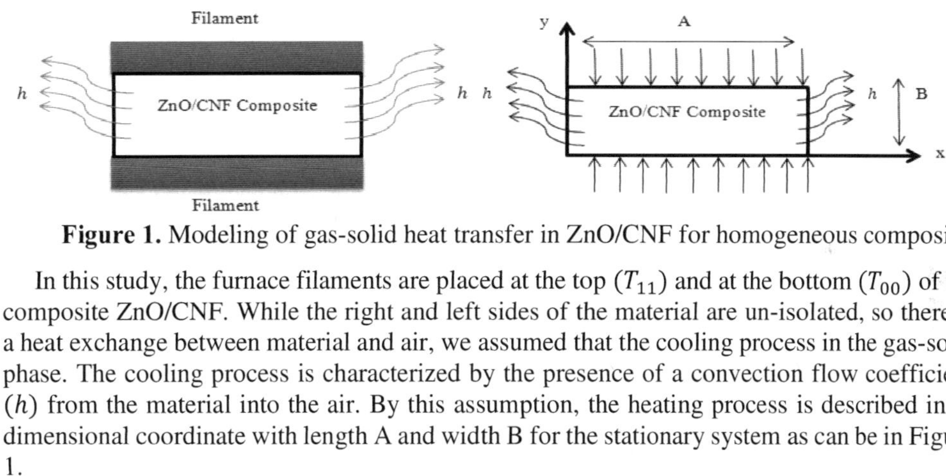


**Figure 1.** Modeling of gas-solid heat transfer in ZnO/CNF for homogeneous composite

In this study, the furnace filaments are placed at the top ($T_{11}$) and at the bottom ($T_{00}$) of the composite ZnO/CNF. While the right and left sides of the material are un-isolated, so there is a heat exchange between material and air, we assumed that the cooling process in the gas-solid phase. The cooling process is characterized by the presence of a convection flow coefficient ($h$) from the material into the air. By this assumption, the heating process is described in 2-dimensional coordinate with length A and width B for the stationary system as can be in Figure 1.

### *2.2 Mathematical Description of Heat Transfer Process*

In this case, a mathematical equation is needed for represents the temperature distribution in the material. For this reason, the diffusion equation can represent the state of the case by the second-order differential equation in 2-dimensional form. The diffusion equation is expressed by:

$$\rho C_p \left(\frac{\partial T}{\partial \tau}\right) = \lambda \left(\frac{\partial^2 T}{\partial x^2} + \frac{\partial T}{\partial y^2}\right) \quad (1)$$

The form of the equation in the left side is the temperature variation of the composite ZnO/CNF with a certain time and in the right side are heat diffusion processes in the composite.

IOP Conf. Series: Materials Science and Engineering **593** (2019) 012016 doi:10.1088/1757-899X/593/1/012016

In the stationary state, there is a heat exchange between composite ZnO/CNF with the environment due to the both sides of the furnace model is un-isolated which was allows the convective heat transfer. Assuming the gas-solid phase heat coefficient is $(h)$ and certain surface area is (a), then the amount of convective heat transfer can be calculated by using the equation $ha(T_2 - T_1)$, where $T_2$ is the temperature in the material, while $T_1$ is the temperature at a boundary between material and the environment. By considering the conduction heat transfer accompanied by convection, the equation (1) becomes the equation (2) as follows:

$$\rho C_p \left(\frac{\partial T}{\partial \tau}\right) = \lambda \left(\frac{\partial^2 T}{\partial x^2} + \frac{\partial T}{\partial y^2}\right) - ha(T_2 - T_1) \qquad (2)$$

The $C_p$is the heat capacity; $\rho$ is the density; $\lambda$ is the thermal conductivity of the composite.

By using the first thermodynamic law for the gas-solid phase, the equation (2) can be solved using the Finite Element Method (FEM). The sources of heat are filaments located at the top and the bottom of composite with continue warm up and heating to the desired temperature.

In the heat transfer simulations that are accompanied by cooling, there are boundary conditions for finding analytical solutions according to the assumption of the model. At the top and bottom, there is a continuous heat source of magnitude $T_{00}$ and $T_{11}$. For the right and left position, there is the heat exchange between the material and the environment. The boundary conditions can be written as follows;

$$\begin{cases} T(x,B) = T_{11}\,;0 < x < A \\ T(x,0) = T_{00}\,;0 < y < B \\ \frac{\partial T}{\partial y}(0,B) = T_{22} \\ \frac{\partial T}{\partial x}(A,y) = 0 \end{cases} \qquad (3)$$

From the boundary conditions applied to the heating system, we can analyze by using the numerical solutions.

*2.3 Discretization of Differential Equations and Boundary Conditions*

The numerical methods used to solve differential equation solutions with certain boundary conditions are by using finite element method. The first is determining domain which can be represented by the global matrix. Furthermore, domain is discretized by dots to form a rectangular which called a grid. These points are represented by the local matrix for determine domain of heat properties. So that material domain is a set of grids that called mesh as can be seen in Figure 2.

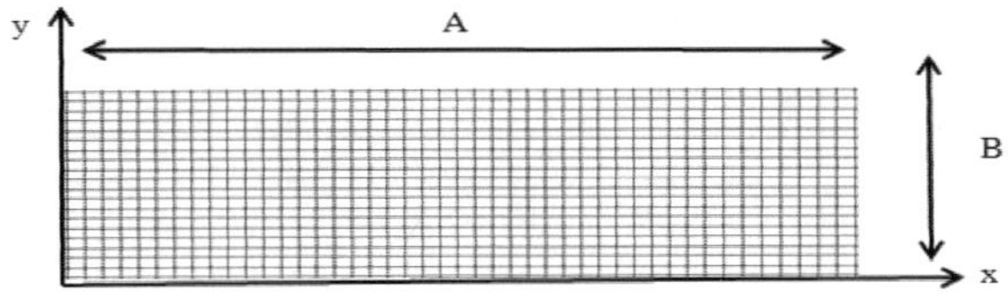


**Figure 2.** Domains and mesh for numerical computing

For example, the domain of material consisting of 50 x 100 grids, representation of the domain is illustrated by the global matrix as shown in Figure 3a and 3b. For illustration of one element is representation by a local matrix as shown in Figure 3c, which is for each corner filled by local number points of each element.

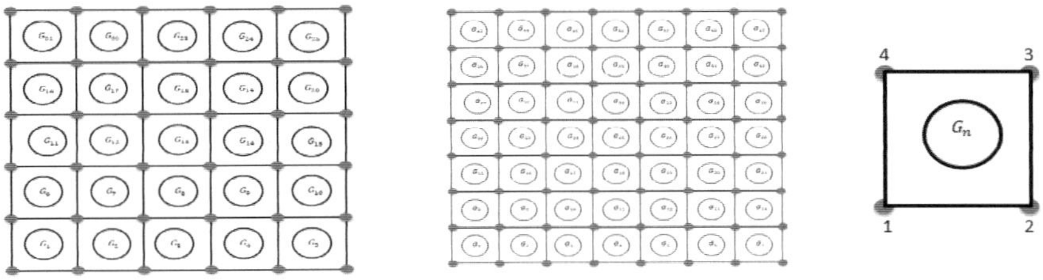


(a) (b) (c)

**Figure 3.** (a) Numerical Computing domains with grids (6x6). (b) for grids (8x8) (c) Element with local point.

After discretizing the domain, a trial function is obtained by;

$$T(x,t) = N_1(x)T_1(t) + N_2(x)T_2(t) + \cdots \quad (4)$$

where $N(x)$ is an interpolation function at the four vertices as expressed:

$$N_1(r,s) = \frac{1}{4}(1-r)(1-s) \qquad N_3(r,s) = \frac{1}{4}(1+r)(1+s)$$

$$N_3(r,s) = \frac{1}{4}(1+r)(1+s) \qquad N_4(r,s) = \frac{1}{4}(1-r)(1+s)$$

The obtained trial function was included in the control differential equation by using Jacobian iterations. It was found that, the heat distribution in each grid changes by the increasing time of annealing.

*2.4 Flow Chart for Numerical Solution*

The flow chart for numerical solutions in this study with differential equation as shown in fig. 4 as follows ;

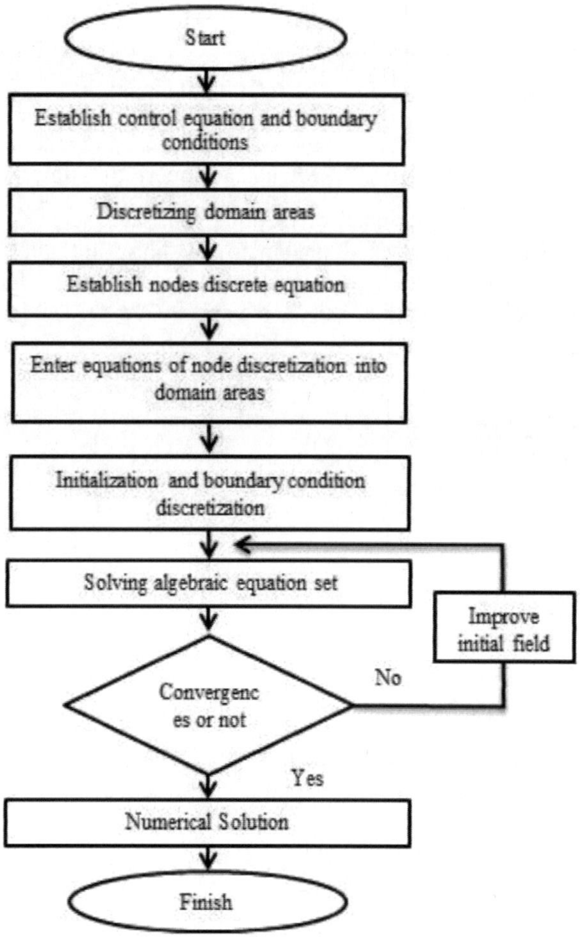


## 3. Result and Discussion

This model uses Jacobian iteration to solve the solution of the differential diffusion equation. The software used is Scilab 6.0.1 from Scilab Enterprises to determine and check the solution is converged or not. The maximum absolute error of two successive iterations is must be below the threshold $10^{-4}$ [22].

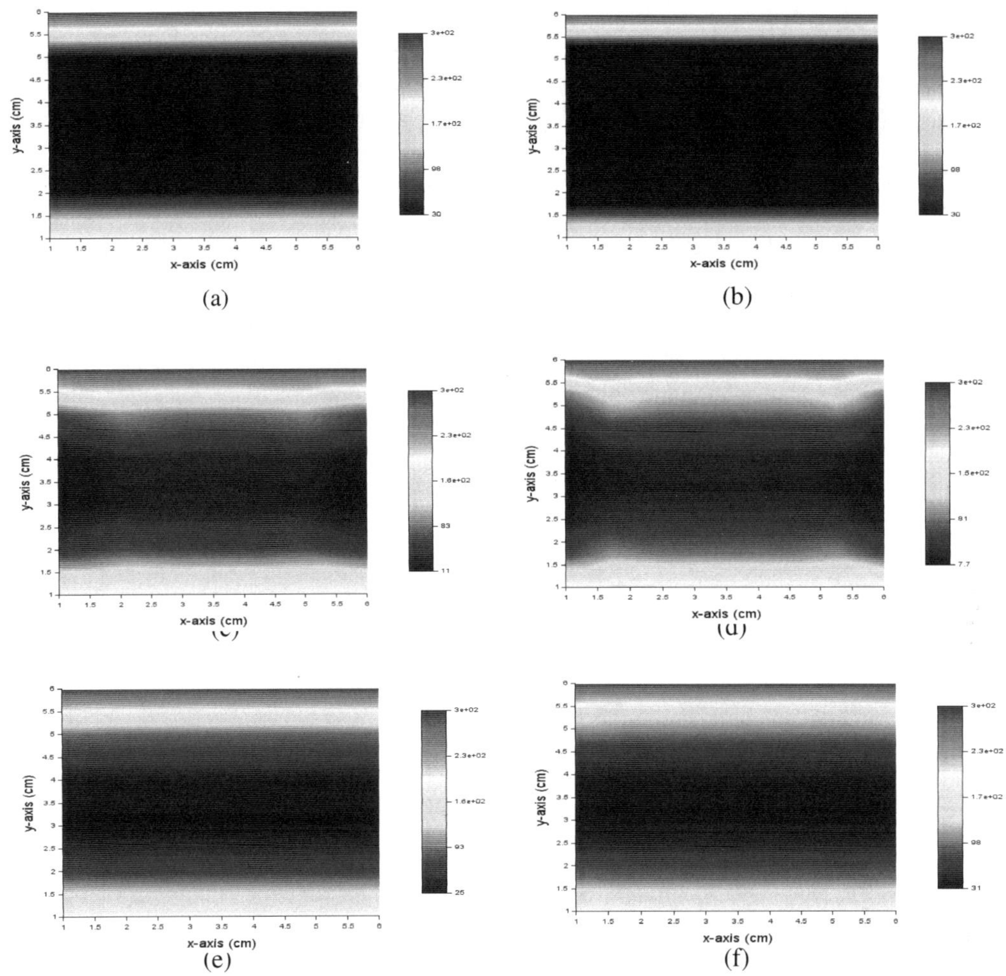


**Figure 5.** State of the plate at the initial temperature (room temperature) (a) $\delta = 0{,}167\ cm$ dan (b) $\delta = 0{,}125\ cm$. Heat distribution asymptotically (c) $\delta = 0{,}167\ cm$ dan (d) $\delta = 0{,}125\ cm$. Heat distribution at steady state (e) $\delta = 0{,}167\ cm$ dan (f) $\delta = 0{,}125\ cm$.

Fig. 5 clearly shows the distribution of temperature for the homogeneous sample inside the furnace. The position of the sample in every condition (from initial to steady state) should be in the middle of the furnace to receive the homogenous temperature during annealing processes.

### *3.1. Verification in grid arrangement*

In the boundary element method, the matrix representation as the domain form which was described the heat distribution for each grid. In this study, the grid arrangement is very important to determine the accuracy of the numerical solutions obtained. The accuracy of the solutions indicated by the number of grids, for high number the solutions are more accurate and approaching the exact solution [22].

For validation in this simulation, the domain of the ZnO/CNF composite is arranged for length and width is 1 cm. Grid were settings by given with variations are 0,167 cm and 0,125 cm with the time iteration for each grid is 2000 steps. The heat distribution for each grid is presented in Figure 3. The solution that approached accuracy results with the numerical solution is for grid 0,125 cm as similar reported in Ref. [22].

### *3.2. Numerical Analysis for ZnO/CNF Plate Composite Heating Process*

Numerical analysis of a differential equation is very important for the accuracy of the solutions obtained. The key point in the numerical analysis is the condition of convergence and stability of asymptotic values. A mathematical model approaches the exact value when the model satisfies the convergence condition. This means that the variations of annealing temperature depend on the grid and the asymptotic state will converge at one particular point. The asymptotic state is constant indicated that the stable value of annealing temperature. In this study shows the asymptotic state for the threshold below $10^{-4}$ is at 25 minute and the temperature at 200 K and 300 K.

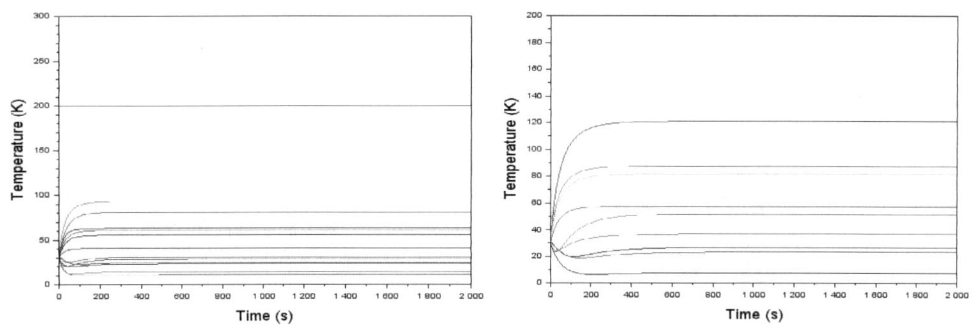


(a) (b)

**Figure 6.** Convergence and stability value from heat distribution at certain grid at (a) $\delta = 0{,}167\ cm$ dan (b) $\delta = 0{,}125\ cm$

Figure 6 shows the left side reduce temperature probably due to the exchange of temperature with air by the convection process and the conduction process in the materials. The effective heat transfer coefficient between conduction process and the convection by the air is calculated by using:

$$k = \frac{1}{\frac{1}{h} + \frac{\phi r}{\lambda}} \tag{5}$$

where $r$ is the radius of particle; $\phi$ is the composite correction factor, equal 0,25.

Figure 7 shows comparison between the steady-state and unsteady-state to the y-axis discretization point. The error value for the steady-state and unsteady-state conditions for the grid 0.125 cm is smaller than that of the grid 0.167 m due to the accuracy solution from FEM is in terms of convergence the number of grids. There are two methods for accurate solution approaches is *h-refinement* which is refers to the process of increasing the number of elements used in the model of particular domain and reduce the size of individual elements. For the second method is *p-refinement* which is emphasizes in the increasing the order of polynomials for an interpolation function [23].

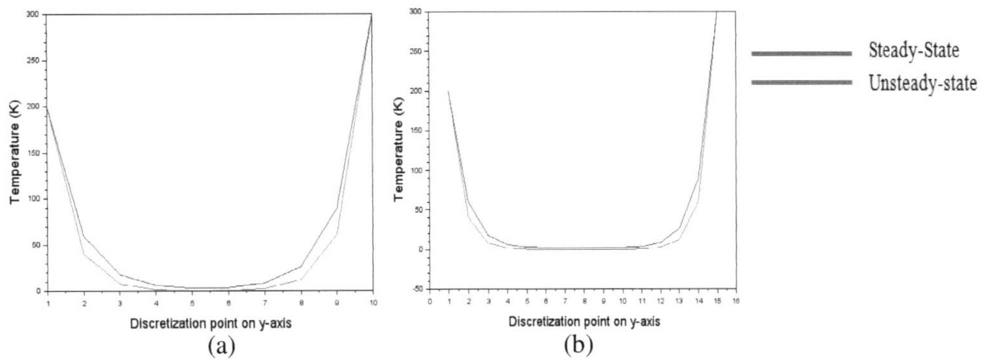


**Figure 7.** Comparison of steady-state and unsteady-state states at the y-axis discretization point (a) $\delta = 0{,}167\ cm$ (b) $\delta = 0{,}125\ cm$.

## 4. Conclusion

A mathematical model for annealing homogenous composite ZnO/CNF is based on the diffusion differential equations for steady-state conditions and dependent on time for stationary domain systems. Numerical solutions for the temperature distribution were obtained by using the finite element method (FEM) which is approach to the exact solution. The solutions can be seen in terms of convergence and stability of the asymptotic values which were analyzed by numerical systems. The efficiency of the annealing system obtained for the asymptotic state is in step 1500 for all grids. Thus, the effective time for annealing composite ZnO/CNF to receive homogenous temperature is 25 minute with a temperature at 200 K and 300 K. The position of the sample in every condition (from initial to steady state) should be in the middle of the furnace to receive the homogenous temperature during annealing processes.

## References

[1]. C H Kim, Bo-Hye Kim 2015 Zinc oxide/activated carbon nanofiber composites for high-performance supercapacitor electrodes *Journal of Power Sources* **274** pp 512–520
[2]. Kai-Ping Wang, Hsisheng Teng 2006 The performance of electric double layer capacitors using particulate porous carbons derived from PAN fiber and phenol-formaldehyde resin. *Carbon* **44** (15) pp 3218–3225
[3]. W Li, F Zhang, Y Dou, Z Wu, H Liu, X Qian, D Gu, Y Xia, B Tu, D Zhao A 2011 self-template strategy for the synthesis of mesoporous carbon nanofibers as advanced supercapacitor electrodes *Advanced Energy Mater* **1** (3) pp 382–386
[4]. Z Ryu, J Zheng, M Wang 1998 Porous structure of PAN-based activated carbon fibers *Carbon* **36** (4) pp 427–432
[5]. L Marcinauskas, Z Kavaliauskas, V Valincius 2012 Carbon and nickel oxide/carbon composites as electrodes for supercapacitors. *Journal of Material science and Technology* **28** (10) pp 931–936
[6]. A Pandolfo, A F Hollenkamp, 2006 Carbon properties and their role in supercapacitors *Journal of Power Sources* **157** (1) pp 11–27

[7]. Y Zhang, H Feng, X Wu, L Wang, A Zhang, T Xia, H Dong, X Li, L Zhang 2009 Progress of electrochemical capacitor electrode materials: A review. *Int. Journal of Hydrogen Energy* **34** (11) pp 4889–4899

[8]. H Wang, Q Gao, J Hu 2010 Preparation of porous doped carbons and the high performance in electrochemical capacitors *Microporous and Mesoporous Materials* **131** (1-3) pp 89–96

[9]. C Largeot, C Portet, J Chmiola, P L Taberna, Y Gogosti, P Simon Relation between the ion size and pore size for an electric double-layer capacitor *Journal of American Chemical Society* **130** (9) pp 2730–2731

[10]. J Hu, H Wang, X Huang 2012 Improved electrochemical performance of hierarchical porous carbon/polyaniline composites *Electrochimica Acta* **74** pp 98–104

[11]. Gyu-Chul Yi, Chunrui Wang and Won Il Park 2005 ZnO nanorods: synthesis, characterization, and application. *Semiconductor Science and Technology* **20** (4) pp S22–S34

[12]. F Lu, W Cai, Y Zhang 2008 ZnO hierarchical micro/nanoarchitectures: solvothermal synthesis and structurally enhanced photocatalytic performance. *Advanced Functional Materials* **18** (7) pp 1047–1056

[13]. D Kalpana, K S Omkumar, S S Kumar, N G Renganathan 2006 A novel high power symmetric ZnO/carbon aerogel composite electrode for electrochemical supercapacitor. *Electrochimica Acta* **52** (3) pp 1309–1315

[14]. M Selvakumar, D K Bhat, A M Aggarwal, S P Iyer, G Sravani 2010 Nano ZnO- activated carbon composite electrodes for supercapacitors *Physica B: Condensed Matter* **405** (9) pp 2286–2289

[15]. L S Aravindaa, K K Nagarajab, H S Nagarajab, K U Bhat, B R Bhat 2013 ZnO/carbon nanotube nanocomposite for high energy density supercapacitors *Electrochimica Acta* **95** pp 119–124

[16]. Y Zhang, X Sun, L Pan, H Li, Z Sun, C Sun, B K Tay 2009 Carbon nanotube-ZnO nanocomposite electrodes for supercapacitors. *Solid State Ionics* **180** (32-35) pp 1525–1528

[17]. G-H Yuan, Z-H Jiang, A Aramata, Y-Z Gao Electrochemical behavior of activated-carbon capacitor material loaded with nickel oxide *Carbon* **43** (14) pp 2913–2917

[18]. K-W Nam, K-H Kim, E-S Lee, W-S Yoon, X-Q Yang, K-B Kim 2008 Pseudocapacitive properties of electrochemically prepared nickel oxides on 3-dimensional carbon nanotube film substrates *Journal of Power Sources* **182** (2) pp 642–652

[19]. T Battumur, S B Ambade, R B Ambade, P Pokharel, D S Lee, S-H Han, W Lee, S-H Lee 2013 Addition of multiwalled carbon nanotube and graphene nanosheet in cobalt oxide film for enhancement of capacitance in electrochemical capacitors *Current Applied Physics* **13** (1) pp 196–204

[20]. Z-S Wu, W Ren, D-W Wang, F Li, B Liu, H-M Cheng High-energy $MnO_2$ nanowire/graphene and graphene asymmetric electrochemical capacitors *ACS Nano* **4** (10) pp 5835–5842

[21]. Z-S Wu, D-W Wang, W Ren, J Zhao, G Zhou, F Li, H-M Cheng 2010 Anchoring hydrous $RuO_2$ on graphene sheets for high-performance electrochemical capacitors *Advanced Functional Materials* **20** (20) pp 3595–3602

[22]. M Ghattassi, J R Roche, F Asllanaj, M Boutayeb Galerkin 2016 method for solving combined radiative and conductive heat transfer *International Journal of Thermal Science* **102** pp 122–136
[23]. Hutton, David D 2004 *Fundamentasl of Finite Element Analysis* (New York: McGraw-Hill)

IOP Conf. Series: Materials Science and Engineering **593** (2019) 012017 doi:10.1088/1757-899X/593/1/012017

# Spatial Analysis For *Pinus merkusii* Land Suitability At Agroforestry Land In Telagah Village Sumatera Utara Indonesia

**Rahmawaty[1*], S Frastika[1], A Rauf[2]**

[1]Department of Forest Management, Faculty of Forestry, Universitas Sumatera Utara, Jl. Tridharma Ujung No.1, Kampus USU Medan, North Sumatra, Indonesia 20155

[2]Faculty of Agriculture, Universitas Sumatera Utara, Medan, Sumatera Utara Indonesia 20155

*Email: rahma2107usu@gmail.com

**Abstract.** Pinus (*Pinus merkusii*) is commonly found in North Sumatra Forest. Community of Telagah Village, Sei Binge Sub District, Langkat, North Sumatera, generally manage the land with agroforestry system. Pinus is the one of agroforestry constituent in the community land in Telagah Village. This study aimed to analyse the suitability of the land for pinus in Telagah Village Sei Bingai Sub District, Langkat District, North Sumatra, Indonesia. This research was conducted on May to September 2018. The survey method was conducted to collect soil samples in the field. Spasial analysis for pinus land suitability was evaluated based on the matching method. The reference and criteria were adopted from the Land Suitability for Agricultural Plants by the Centre for Soil and Agroclimate Research, Bogor-Indonesia. The Geographic Information System (GIS) was used to map the actual and potential land suitability in Telagah Village. The results showed that the actual land suitability classes for pinus in this area were moderately suitable (S2) (97.26%) and marginal suitable (S3) (2.74%). The potential land suitability classes for pinus in this area were Suitable (S1) (52.04), moderately suitable (S2) (46.04%) and marginal suitable (S3) (1.92%) in Telagah Village. The temperature (tc) and erosion hazard (eh) were the dominant of limiting factor in this area.

## 1. Introduction

Pinus (*Pinus merkusii*) is classified in Pinaceae family, which has various local names such as: tusam (Indonesian), uyam (Aceh), son song bai (Thai), mercus pine (trade), mindoro pine (Philippines), and tenasserim pine (Inggris). the selection of this species is preferred because of the availability of seeds, rapid growth rate, can be a pioneer, and can grow on marginal lands [1, 2). Land suitability evaluation needs to be done to be a basis for consideration in land use decision making [3]. Land evaluation is a process of assessing the potential of a land for certain uses. Land suitability is the suitability of a land for certain uses [4, 5, 6, 7]. Land evaluation results are used as a basis for rational land use planning, so that land can be used optimally and sustainably [8]. Land suitability classes into two namely: actual and potential land suitability class [5, 9, 10]. The process of land evaluation is comparing the requirements requested by the type of land use that will be applied to the properties of the land that is owned by the land to be used.

Geographical Information System (GIS) is a tool that can be apply for land suitability. One application of GIS is mapping the results of land evaluation and presenting the results in the form of maps to show the spatial distribution of geographical phenomena including characteristics that are in accordance with those on the surface of the earth [2, 11, 12, 13, 14]. The GIS also has been widely applied in various fields Land evaluation results can be described in the form of maps using GIS technology. It is used for data collection, storage, analysis and manipulation of geographic references.

Pinus can be found in several places, including on community lands, such as in agroforestry land. One place that has the potential to be found in Telagah Village, Sei Bingai Sub District, Langkat, North Sumatra, Indonesia. Research on mapping of land suitability for Pinus as one of agroforestry plants in this location has never been done. Hence, this study aimed to to analyse the suitability of the land for pinus in Telagah Village Sei Bingai Sub District, Langkat District, North Sumatra, Indonesia.

## 2. Materials and Methods

This research was conducted on May to September 2018 in Telagah Village, Sei Bingai Sub District, Langkat, North Sumatra, Indonesia (Figure 1). The survey method was conducted to collect soil samples in the field. The tools used in this research were: abney level, altimeter, label, scales, plastic bag, hoe, Haga hypsometer, compass, camera, stationery, and tally sheet. Land suitability classification (LSC) was evaluated based on the matching method Ritung *et al* [16]. The reference and criteria were adopted from the Land Suitability for Agricultural Plants by the Centre for Soil and Agroclimate Research, Bogor-Indonesia [8]. The SIG was used to analyze the spatial distribution of land suitability classes, both actual and potestial.

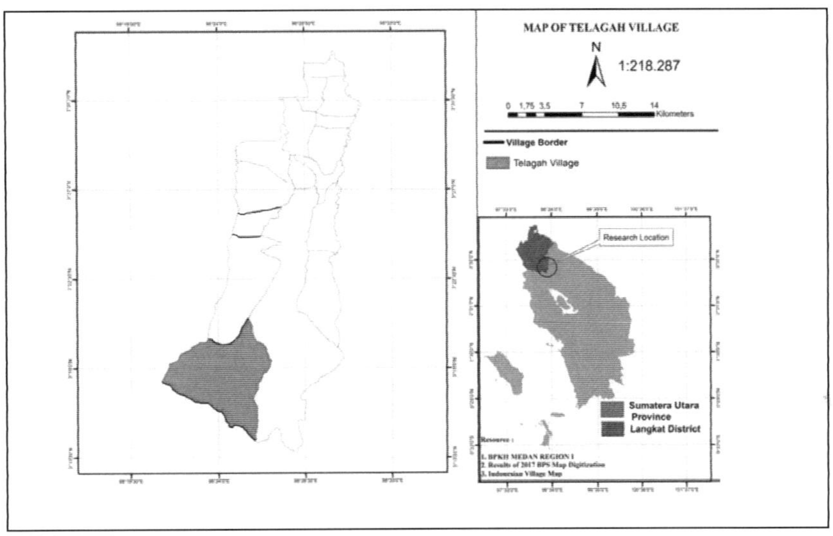


**Figure 1.** Map of Telagah Village, Sei Bingai Sub District, Langkat, North Sumatra, Indonesia

The results of the assessment of the land suitability classes were presented in the form of tables and maps that provide a class description of the land suitability of pinus. The assessment and presentation of land suitability class results were based on FAO [16]. There are four classes, namely: Highly suitable (S1), Moderately suitable (S2), marginal suitable (S3), and not suitable (N). The same method has been used in analyzing the suitability of potato in Siosar [14]; in analyzing the suitability of oil palm [19], and in analyzing the suitability of trees in Besitang Watershed [1]; analyzing the suitability of Jati in Arboretum USU [6] and in analyzing the suitability of alpukat and durian in Karo District [17].

## 3. Results and discussion

The Land Unit Area in Telagah Village Sei Bingai District was presented in Table 2 and delineated in Figure 1 and Figure 2.

**Table 1.** Land Unit Area in Telagah Village

| Land Unit | Area | |
|---|---|---|
| | Ha | % |
| I | 2.422,70 | 52,04 |
| II | 206,39 | 4,43 |
| III | 38,22 | 0,82 |
| IV | 99,34 | 2,13 |
| V | 1.799,47 | 38,65 |
| VI | 89,33 | 1,92 |
| Total | 4.655,45 | 100,00 |

The results showed that the land unit consists of 6 classes (Table 1) with different areas. The widest land unit was land unit I covering approximately 2.422,70 ha (52,04% from total area), followed by land unit 5 covering approximately 1.799,47 ha (38,65% from total area).

The land suitability classes for Pinus were moderately suitable (S2) and marginal suitable (S3) in Telagah Village (Tabel 2). There were suitable (S1) found at agroforestry land in this village based on analyse of potential land suitability. There were no not suitable (N) classes were found at agroforestry land in this village. Class S3 means that lands having limitations, which are severe for sustained application of a given use and will so reduce productivity or benefits or increase required inputs that this expenditure will be only marginally justified [16, 19]. There were several limiting factors in land suitability evaluation in the area, namely: the availability of water (wa), erosion hazard (eh), and root zone medium (rc), Oxigen avaibility(oa) and nutrient retention (nr). The water availability (wa) was the dominant limiting factor in this area. Land suitability classification (actual and potential land suitability class of Pinus) in Telagah Village, Sei Bingai Sub District was presented in Table 2 and Table 3 and delineated in Figure 2 and Figure 3.

**Table 2.** The actual land suitability classes for *Pinus merkusii* in Telagah Village

| Actual Land Suitability | Area | |
|---|---|---|
| | Ha | % |
| S2,nr | 2,422.70 | 52.04 |
| S2,tc,eh | 1,898.80 | 40.79 |
| S2,tc,nr | 206.39 | 4.43 |
| S3,eh | 38.2 | 0.82 |
| S3,rc | 89,33 | 1.92 |
| Total | 4,655.45 | 100.00 |

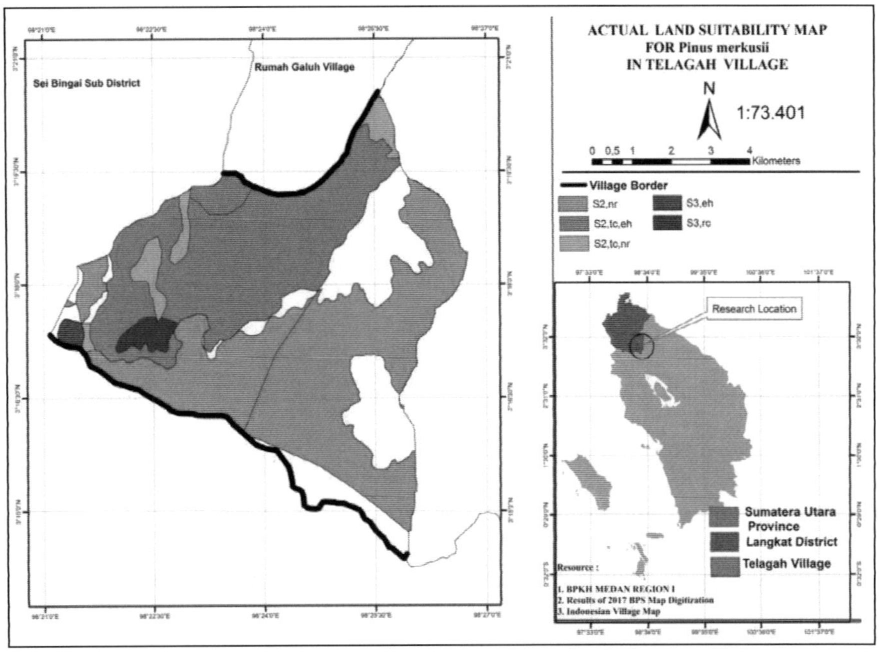


**Figure 2.** Map of actual land suitability in Telagah Village

**Table 3.** The potential land suitability classes for *Pinus merkusii* in Telagah Village

| Potential Land Suitability | Area | |
|---|---|---|
| | Ha | % |
| S1 | 2,422.70 | 52.04 |
| S2,tc | 2,105.19 | 45.22 |
| S2,tc,eh | 38.2 | 0.82 |
| S3,rc | 89.33 | 1.92 |
| Total | 4,655.45 | 100.00 |

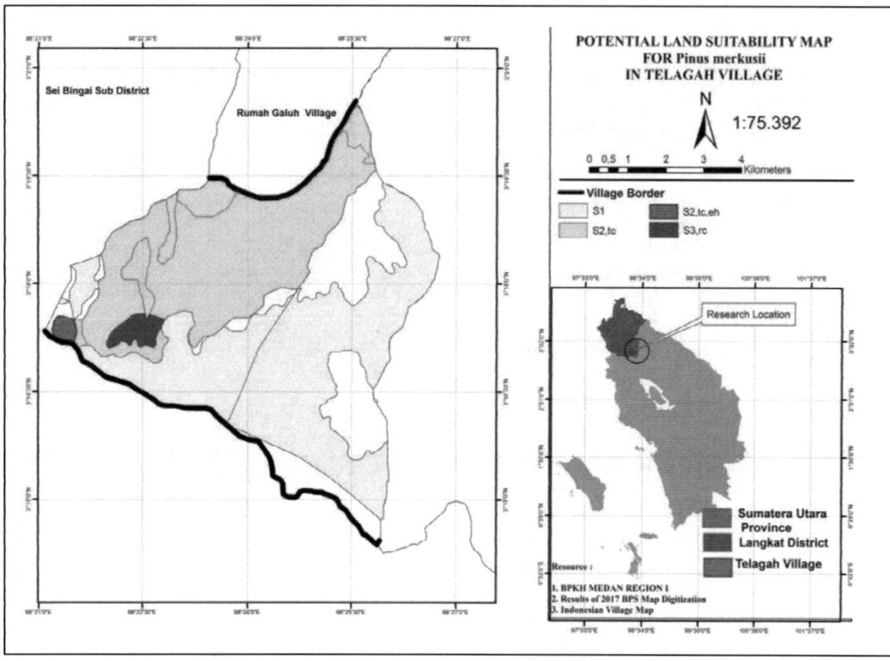


**Figure 3.** Map of potential land suitability in Telagah Village

Based on Table 2, Tabel 3, Figure 2 dan Figure 3, in this area were moderately suitable (S2) (97.26%). It means that in Telagah village, *Pinus merkusii* was moderately suitable to be developed. *Pinus merkusii* was preferred because it can increase people's income. Wood and pine sap can be sold easily so that it can increase people's income. *Pinus merkusii* is a gondorukem-producing tree. Gondorukem is used to mix ingredients such as: batik makers, soap, paint and varnish, paper, while turpentine is used for oil paint, a mixture of perfumes, detergents, flavoring agents, protective coating, insecticides, lubricants, medicine, plastic, rubber.

## 4. Conclusions

The actual land suitability classes for Pinus in this area were moderately suitable (S2) (97.26%) and marginal suitable (S3) (2.74%). The potential land suitability classes for pinus in this area were Suitable (S1) (52.04), moderately suitable (S2) (46.04%) and marginal suitable (S3) (1.92%) in Telagah Village. The temperature (tc) and erosion hazard (eh) were the dominant of limiting factor in this area.

**References**

[1] Mangundikoro A 1983 *Pattern and Strategy for Developing Forests of Pinus merkusii-in Bahasa* (Jakarta: Simpo Pinus'83 Proceeding)

[2] Rahmawaty, Sembiring I E P, Batubara R, and Patana P 2018 Mapping of Tree Damage Classification in the Western Part of Medan City Green Belts Using Geographic Information System (IOP Conference Series: Earth and Environmental Science)

[3] Ritung S, Wahyunto, Agus F and Hidayat H 2007 Land Suitability Evaluation Guidence. Balai Penelitian Tanah dan *World Agroforestry Centre*. Bogor.

[4] Azis A, Sunarminto BH and Renanti MD 2005 Land Suitability Evaluation for Cultivation of Food Crops Using Artificial Neural Networks (Yogyakarta: Universitas Gajah Mada)

[5] Arsyad S 2010 *Soil and Water Conservation (Konservasi Tanah dan Air) 2nd Edition* (Bogor: IPB Press) p 472

[6] Rahmawaty, Siregar N C and Rauf A 2016 Land Suitability For Tectona Grandis;"Case Study in Arboretum Kwala Bekala, Sumatera Utara University (*Kesesuaian Lahan Tanaman Jati; "Studi Kasus di Arboretum Kwala Bekala, Universitas Sumatera Utara*). *Jurnal Penelitian Ekosistem Dipterocarpa* **2**(2)

[7] Rahmawaty, Villanueva TR and Carandang MG 2011 Participatory Land Use Allocation, Case Study in Besitang Watershed, Langkat, North Sumatera, Indonesia. (Jerman: Lambert Academic Publishing) p 200

[8] Hardjowigeno S and Widiatmaka 2007 Land Suitability Evaluation and Land Use Planning. (Yogyakarta: Gajah Mada University Press)

[9] Djaenudin D, Hendrisman M, Subagjo H and Hidayat A 2003 Technical Guidelines for Evaluating Land for Agricultural Commodities. (Bogor: Pusat Penelitian dan Pengembangan Tanah dan Agroklimat)

[10] Hardjowigeno S 1994 *Land Suitability for Agricultural Development, Recreation and Building Areas- in Bahasa* (Bogor: Bogor Agricultural University) p 381

[11] Rahmawaty, Sitorus NA and Rauf A 2017a Distribution, above-ground biomass and carbon stock of the vegetation in Taman Beringin Urban Forest, Medan City, North Sumatra, Indonesia. *Malaysian Forester* **80** (1)

[12] Rahmawaty, Patana P and Latifah S 2017b Spatial analysis on distribution of green belt to reduce impacts of climate change in Medan City, North Sumatra. *Malays. Appl. Biol.* **46**(2) pp 67–76

[13] Satriawan H, Harahap E M, Rahmawaty and Karim A 2015 Effectiveness of Soil Conservation to Erosion Control on Several Land Use Types. *Agriculture Poľnohospodárstvo)* **61** (2) pp 61–68

[14] Tarigan A, Rauf A and Rahmawaty 2016 Land Suitability Evaluation for Potato in the Karo District Siosar Relocation Area-in Bahasa *Jurnal Pertanian Tropik* **3**(2) (14) pp 124–131

[15] Ritung S, Nugroho K, Mulyani A and Suryani E 2011 Technical guidelines for evaluating land for agricultural commodities (Revised edition)-in Bahasa. (Bogor: Badan Penelitian dan Pengembangan Pertanian) p168

[16] Food and Agriculture Organization (FAO) 1976 *A Framework for Land Evaluation* (Rome, Italy: Soil Bulletin 32) p 72

[17] Rahmawaty, Elsi Kurnia Sari, Ahmad Syofyan, Abdul Rauf 2015 Integrated Geographic Information System and Global Positioning System for Mapping of Forest Plants in Supporting Natural Resources Protection *Procedia Chemistry* **14** pp 334–342

[18] Soenardi 1983 Soenardi 1983 *Marketing of Pine Wood and its results*. (Jakarta: Proceeding Simpo Pinus'83)

**Acknowledgments**

We are grateful to the Organizing Committee of the 14th Pacific RIM BIO-BASED Composite Symposium 2018 for give us an opportunity to present of this work on 30 October 2018, at the Swiss Bell Hotel, Makassar, as well as, University of Sumatera Utara, the students of Department of Forest Management, Faculty of Forestry, USU who have conducted research in Sei Bingai, Langkat, North Sumatera Province, Indonesia.

IOP Conf. Series: Materials Science and Engineering **593** (2019) 012018 doi:10.1088/1757-899X/593/1/012018

# Effect of $CO_2$ on the degradation of cement-bonded particleboard based on the treatment to conventional cured board

**R S Maail**

Department of Forest Products Technology, Faculty of Agriculture Pattimura University,
Jl. Ir. M. Putuhena, Ambon, Indonesia

e-mail : rohny_maail@yahoo.com

**Abstract.** This study aimed to evaluate and clarify the effect of $CO_2$ at either the gaseous or supercritical phase on the degradation of cement-bonded particleboard (CBP) based on the treatment to conventional cured board, while the estimation of long term degradation of CBP is also discussed. The properties of CBP are improved by $CO_2$ treatment in short period in both the gaseous and supercritical phases even after the conventional curing process. These properties, however, over the longer treatment time degraded in rapid rate against $CO_2$ treatment in both the gaseous and supercritical phases. It was considered that the degradation of CBP increased with increasing the $CO_2$ concentration, $CO_2$ pressure and treatment time. With the longer treatment time against supercritical $CO_2$ in higher concentration and pressure, the residual values of mechanical properties (IB strength, MOR, and MOE) were decreased, indicating CBP performance was degraded rapidly compared to treatment in the lower concentration and pressure of $CO_2$ or under gaseous phase. High coefficient determination values between the residual values of CBP properties, and the treatment time and $CO_2$ concentration were observed and the simple linear model was found to estimate the long term degradation of CBP among the properties and affecting factors.

## 1. Introduction

A well-known phenomenon associated with cement-based materials placed in natural environmental locations is the carbonation that occurs naturally over numerous years owing to the reaction with atmospheric $CO_2$.[1] The carbonation process occurs rather slowly in natural environment due to the low concentration of $CO_2$ in the environment which is only 0.03-0.04% by volume. Most of the work carried out on the carbonation of cementation materials have been concentrated in the variation of the microstructure and mechanical properties in contact with $CO_2$ as a function of the type of cement being used for the concrete or cement composites. Different authors[2-6] have investigated the effects of carbonation on pore structure, changes of specific surface area, permeability, and diffusion properties of hydrated cement pastes and mortars as the degradation process. The variations of different properties i.e., the permeability, porosity, and pore size distribution as a consequence of the carbonation process in cement-based materials [5-7]. Additionally a few comparative studies on carbonation have been carried out on cement samples, resulting that microstructure of cement paste carbonated using supercritical $CO_2$ was different than that of natural-carbonated paste. The depth of carbonation in real environmental condition was estimated based on the result obtained from the accelerated carbonation test on the fly ash and blast-furnace slag concrete [8].

IOP Conf. Series: Materials Science and Engineering **593** (2019) 012018 doi:10.1088/1757-899X/593/1/012018

The carbonation is also one of the major causes of degradation of cement-bonded particleboard (CBP) over the time due to contact with $CO_2$. This phenomenon was clarified in our previous study,[9] in which $CO_2$ in a supercritical phase accelerates the curing process and enhances the mechanical properties of CBP in the short treatment time, however, in the longer time span leads to degradation in mechanical properties because of the carbonation of cement, resulting negative effect on board performance.

In the experimental works, data about the long-term performance of CBP are scarce. Therefore, in the present study, it is desirable to accelerate the carbonation process to shorten the duration of treatment time to predict long term performance of CBP. The acceleration can be performed either by increasing the concentration or the pressure of $CO_2$. Although many studies on manufacturing CBP using $CO_2$ in either setting or curing process at both the gaseous and supercritical phase have been done,[9-14] none has clarified the estimation of degradation process of CBP based on the result obtained from $CO_2$ treatment to conventional cured board, beside the literatures concerned with the analysis of carbonation which is affected to the degradation of CBP at elevated pressures and concentration of $CO_2$ are less common. It was then possible to observe the degradation of CBP by measuring the residual values of CBP properties based on the degradation values of CBP properties treated at both the gaseous and supercritical CO2 phases in certain period.

The objectives of this study were to evaluate and clarify the effect of CO2 at either the gaseous or supercritical phase on the degradation of CBP based on the treatment to conventional cured board and the results were applied to estimate long term degradation of CBP by applying the equation models from statistical method and analysis.

## 2. Materials and methods

### *2.1 Materials*

Mixtures with proportionally equal amounts of particles of Japanese cypress (*Chamaecyparis obtusa* Endl) and Japanese cedar (*Cryptomeria japonica* D.Don) were used to manufacture CBP. The average size of the wood particles was 3.3 mm (L), 0.7 mm (W), and 0.1 mm (T). Ordinary Portland cement of Osaka Sumitomo Co. Ltd. was used as a binder; $CO_2$ was used as an accelerator for curing and carbonation processes at either the gaseous or supercritical phase. To produce supercritical $CO_2$, $CO_2$ gas was maintained in the liquid phase by passing the gas through a condenser. The liquid $CO_2$ was then pumped into the reaction cell surrounded by a water jacket set at 40°C at a pressure up to 10 MPa.[9]

### *2.2 Manufacture and treatment of CBP*

CBP with a targeted density of 1.2 g/cm$^3$ was manufactured at a cement/wood particle/water weight ratio of 2.5:1.0:1.25. Hand-formed mats of 230 x 230 mm were cold pressed to a targeted thickness of 12 mm and kept in an oven set at 60°C for 24 h. Afterwards, CBP was wrapped with a polyvinylchloride (PVC) film immediately, followed by curing for 28 days at room temperature (conventional). Four specimens of 50 x 210 mm prepared from these boards were then treated with (1) gaseous $CO_2$; and (2) supercritical $CO_2$. In order to estimate long term of CBP degradation, the following two types of experiment were conducted at various levels of $CO_2$ pressures and concentrations for various treatment times. In the first experiment, specimens were subjected in a reaction cell surrounded by a water jacket set at 26°C to $CO_2$ pressure of 1.0 MPa at gaseous phase for 30 min; 3, 5, 10, 24 h and 10 days treatment. The concentrations of $CO_2$ applied were 10, 50 and 100%. In the second experiment, specimens were subjected in a reaction cell surrounded by a water jacket set at 40°C to $CO_2$ pressure of 10 MPa at supercritical phase for 30 min; 3, 5, 10, 24 h and 10 days treatment. The concentrations of $CO_2$ applied were 10, 50 and 100%. Later, specimens of both experiments were dried in an oven drying at 80°C for 10 h, followed by conditioning at ambient temperature prior to property evaluation.

### *2.3 Evaluation of CBP properties*

The mechanical and dimensional properties of the boards were tested in accordance with the Japan Industrial Standard (JIS) A 5908 (1994). The boards were cut into 50 x 210 mm samples for the static bending test and 50 x 50 mm samples for the internal bond (IB) strength, thickness swelling (TS), and water absorption (WA) tests. The static bending test was conducted using a three-point bending test over an effective span of 180 mm (15 times the board thickness) at a loading speed of 10 mm/min. Four test samples were prepared from each treatment group for the foregoing tests, and the average values were calculated.

*2.4 Measurement of residual values of CBP properties*

The residual values of CBP properties (%) which indicates the degradation of CBP properties from the initial values (100%) was measured based on the degradation values of CBP properties treated at both the gaseous and supercritical $CO_2$ in certain period compared to the optimum values (the initial values) of CBP properties treated in 30 min. The residual values of CBP properties were calculated from the following equation;

$$\text{Residual values of properties } (\%) = (P \,/\, P_0) \times 100 \quad (1)$$

where $P$ is the values of CBP properties treated at both the gaseous and supercritical CO2 in certain period; and $P_0$ is the optimum values (the initial values) of CBP properties treated at both the gaseous and supercritical $CO_2$ in 30 min.

*2.5 Statistical Analysis*

To estimate long term degradation of CBP affected by $CO_2$ in gaseous and supercritical phase, the multiple regression analyses were performed with commercially available statistical software (SPSS Statistics 17.0, SPSS Japan Inc.) from the fallowing equation;

$$y = ax_1 + bx_2 + c \quad (2)$$

where $y$ is the variable being predicted (CBP properties); $x_1$, $x_2$ are the predictor variables in the equations, where $x_1$ is the $CO_2$ concentration (10, 50 and 100%) and $x_2$ is the treatment time (30 min, 3, 5, 10 and 24 h, 10 days); $c$ is the $y$-intercept which indicates the point at which the regression plane intersects the y-axis when the values of predictor scores are all zero; a and b are the regression coefficients which are used as multipliers for the corresponding predictor variables (i.e., $x_1$ and $x_2$).

## 3. Results and discussion

*3.1 Effect of carbon dioxide phase, concentration and various treatment times on the properties of CBP*

The average values for the mechanical properties (IB strength, MOR and MOE) and dimensional stabilities (TS and WA) of conventional cured CBP treated by $CO_2$ in gaseous and supercritical phase at various concentrations and treatment times are presented in Table 1, which indicates that the IB strength, MOR, MOE, TS and WA were strongly affected by $CO_2$ treatment in both the gaseous and supercritical $CO_2$ conditions. With the increase in treatment time up to 30 min by the gaseous $CO_2$ at 1.0 MPa of pressure or by the supercritical $CO_2$ at 10 MPa of pressure; the mechanical properties (IB strength, MOR, MOE) and dimensional stabilities (TS and WA) were significantly improved with increasing the $CO_2$ concentration compared to control values. However, in the short treatment time, the optimum properties of the supercritical $CO_2$ cured CBP could achieve the higher values than those of the gaseous $CO_2$. This condition indicated that the gaseous $CO_2$ and the supercritical $CO_2$ treatments improved the properties of CBP even after conventional curing process for 28 days. This finding shows different results compared to the results on previous research which indicated that the optimum properties of CBP manufactured by supercritical $CO_2$ curing were almost similar to those of gaseous

IOP Conf. Series: Materials Science and Engineering **593** (2019) 012018 doi:10.1088/1757-899X/593/1/012018

**Tabel 1**. Effect of carbon dioxide phase, concentration and various treatment times on the properties of CBP

| Experiments | Properties | Unit | $CO_2$ concentration | | | | | | | | | | | | | | | | | | | | |
|---|---|---|---|---|---|---|---|---|---|---|---|---|---|---|---|---|---|---|---|---|---|---|---|
| | | | 10% | | | | | | | 50% | | | | | | | 100% | | | | | | |
| | | | Control | 30 min | 3 h | 5 h | 10 h | 24 h | 10 days | Control | 30 min | 3 h | 5 h | 10 h | 24 h | 10 days | Control | 30 min | 3 h | 5 h | 10 h | 24 h | 10 days |
| 1.0 MPa | IB | MPa | 0.59 | 0.67 | 0.66 | 0.66 | 0.65 | 0.64 | 0.62 | 0.59 | 0.79 | 0.77 | 0.75 | 0.74 | 0.73 | 0.71 | 0.59 | 0.86 | 0.84 | 0.83 | 0.81 | 0.79 | 0.77 |
| (Gaseous | | (STD) | (0.16) | (0.04) | (0.02) | (0.02) | (0.03) | (0.11) | (0.02) | (0.16) | (0.04) | (0.05) | (0.07) | (0.05) | (0.04) | (0.02) | (0.16) | (0.12) | (0.04) | (0.04) | (0.04) | (0.04) | (0.05) |
| $CO_2$-phase) | MOR | MPa | 10.09 | 10.26 | 10.25 | 10.23 | 10.23 | 10.21 | 10.20 | 10.09 | 10.96 | 10.89 | 10.82 | 10.72 | 10.68 | 10.42 | 10.09 | 12.2 | 12.01 | 11.97 | 11.65 | 11.50 | 11.13 |
| | | (STD) | (0.55) | (0.25) | (0.91) | (0.49) | (0.49) | (0.22) | (0.20) | (0.55) | (0.20) | (0.29) | (0.63) | (0.26) | (0.61) | (0.44) | (0.55) | (0.59) | (0.54) | (0.14) | (0.83) | (0.47) | (0.75) |
| | MOE | GPa | 3.49 | 3.61 | 3.54 | 3.53 | 3.48 | 3.43 | 3.42 | 3.49 | 4.18 | 4.04 | 4.01 | 3.97 | 3.90 | 3.83 | 3.49 | 4.82 | 4.72 | 4.67 | 4.46 | 4.35 | 4.15 |
| | | (STD) | (0.74) | (0.38) | (0.39) | (0.38) | (0.20) | (0.24) | (0.14) | (0.74) | (0.50) | (0.24) | (0.18) | (0.31) | (0.27) | (0.26) | (0.74) | (0.25) | (0.55) | (0.32) | (0.60) | (1.01) | (0.66) |
| | TS | % | 0.61 | 0.70 | 0.71 | 0.72 | 0.73 | 0.73 | 0.74 | 0.61 | 0.67 | 0.68 | 0.70 | 0.70 | 0.72 | 0.73 | 0.61 | 0.56 | 0.57 | 0.60 | 0.62 | 0.62 | 0.65 |
| | | (STD) | (0.12) | (0.17) | (0.12) | (0.16) | (0.12) | (0.20) | (0.26) | (0.12) | (0.11) | (0.11) | (0.07) | (0.16) | (0.19) | (0.12) | (0.12) | (0.04) | (0.06) | (0.07) | (0.08) | (0.06) | (0.09) |
| | WA | % | 21.39 | 20.20 | 20.34 | 20.20 | 20.31 | 20.37 | 21.56 | 21.39 | 19.42 | 19.52 | 19.63 | 19.81 | 20.21 | 20.32 | 21.39 | 20.05 | 20.07 | 20.13 | 20.37 | 21.56 | 22.07 |
| | | (STD) | (1.03) | (1.53) | (1.39) | (1.02) | (1.25) | (0.78) | (1.04) | (1.03) | (0.31) | (0.50) | (0.76) | (0.89) | (1.15) | (0.91) | (1.03) | (0.93) | (1.34) | (0.83) | (0.87) | (0.67) | (1.02) |
| 10 MPa | IB | MPa | 0.59 | 0.81 | 0.77 | 0.74 | 0.72 | 0.71 | 0.69 | 0.59 | 1.18 | 1.15 | 1.13 | 1.12 | 1.10 | 1.12 | 0.59 | 1.35 | 1.31 | 1.28 | 1.26 | 1.22 | 1.14 |
| (Supercritical | | (STD) | (0.16) | (0.11) | (0.06) | (0.07) | (0.05) | (0.06) | (0.04) | (0.16) | (0.08) | (0.11) | (0.19) | (0.05) | (0.13) | (0.20) | (0.16) | (0.11) | (0.14) | (0.06) | (0.13) | (0.13) | (0.07) |
| $CO_2$-phase) | MOR | MPa | 10.09 | 11.24 | 11.07 | 11.03 | 10.79 | 10.33 | 10.26 | 10.09 | 12.52 | 11.79 | 11.55 | 11.38 | 11.02 | 10.38 | 10.09 | 14.81 | 13.97 | 13.33 | 12.98 | 12.47 | 11.50 |
| | | (STD) | (0.55) | (0.71) | (0.23) | (0.11) | (0.28) | (0.21) | (0.16) | (0.55) | (0.89) | (0.65) | (0.72) | (0.43) | (0.67) | (0.77) | (0.55) | (0.57) | (0.74) | (0.77) | (0.33) | (0.59) | (0.44) |
| | MOE | GPa | 3.49 | 5.41 | 5.34 | 5.09 | 4.78 | 4.66 | 4.52 | 3.49 | 6.35 | 6.15 | 6.11 | 5.73 | 5.36 | 5.00 | 3.49 | 7.05 | 6.74 | 6.67 | 6.42 | 6.32 | 5.88 |
| | | (STD) | (0.74) | (0.49) | (0.25) | (0.13) | (0.10) | (0.50) | (0.33) | (0.74) | (0.41) | (0.21) | (0.63) | (0.46) | (0.28) | (0.18) | (0.74) | (0.17) | (0.48) | (0.42) | (0.46) | (0.57) | (0.26) |
| | TS | % | 0.61 | 0.57 | 0.57 | 0.60 | 0.62 | 0.63 | 0.64 | 0.61 | 0.54 | 0.54 | 0.56 | 0.56 | 0.59 | 0.61 | 0.61 | 0.52 | 0.53 | 0.56 | 0.58 | 0.61 | 0.62 |
| | | (STD) | (0.12) | (0.29) | (0.20) | (0.14) | (0.12) | (0.22) | (0.11) | (0.12) | (0.10) | (0.12) | (0.09) | (0.15) | (0.20) | (0.13) | (0.12) | (0.09) | (0.12) | (0.08) | (0.16) | (0.12) | (0.09) |
| | WA | % | 21.39 | 18.56 | 18.63 | 18.76 | 18.79 | 18.80 | 18.89 | 21.39 | 19.05 | 19.15 | 19.24 | 19.35 | 19.53 | 20.72 | 21.39 | 19.65 | 19.99 | 20.10 | 20.28 | 20.73 | 22.27 |
| | | (STD) | (1.03) | (1.16) | (1.03) | (1.10) | (1.02) | (1.14) | (1.01) | (1.03) | (1.23) | (1.31) | (1.02) | (1.14) | (1.16) | (1.29) | (1.03) | (1.01) | (1.03) | (0.98) | (1.10) | (1.06) | (1.03) |

*STD* is standard deviation

$CO_2$ curing, beside the addition of carbon dioxide at the gaseous of 0.5 MPa and 1.0 MPa yielded the lowest bonding properties and it showed improvement proportionally at a longer curing time [9]. In contrast, in our present study, it was evident that the optimum properties of CBP treated by supercritical $CO_2$ even after conventional curing process were higher than that by the gaseous $CO_2$, especially at 100% of $CO_2$ concentration. The additions of carbon dioxide at both the gaseous and supercritical $CO_2$-phase at 30 min improved the bending properties and dimensional stability of CBP even after conventional curing process for 28 days. Then, over the treatment from 3 h, these properties were continuously decreased. It should be noticed that the mechanical properties and dimensional stabilities of conventional cured CBP seemed to reach the higher values in the shorter treatment time with increasing of $CO_2$ concentration and pressure condition, however, the extension of treatment time over 3 h to 10 days resulted in any decreases of mechanical properties and dimensional stabilities of these boards, indicating that degradation process was accelerated in both gaseous and supercritical $CO_2$ treatments, as an effect of increasing the concentration and pressure conditions of carbon dioxide. The carbonation process in short period improved the mechanical and dimensional properties of CBP and in longer period degraded those properties.

Since the properties of CBP are improved by $CO_2$ treatment in both the gaseous and supercritical phase even after the conventional curing process, this phenomenon suggests cement will be cured at the initial period but degrade in longer period by the carbonation that occurs naturally over numerous years owing to the reaction with atmospheric $CO_2$ in ambient conditions.

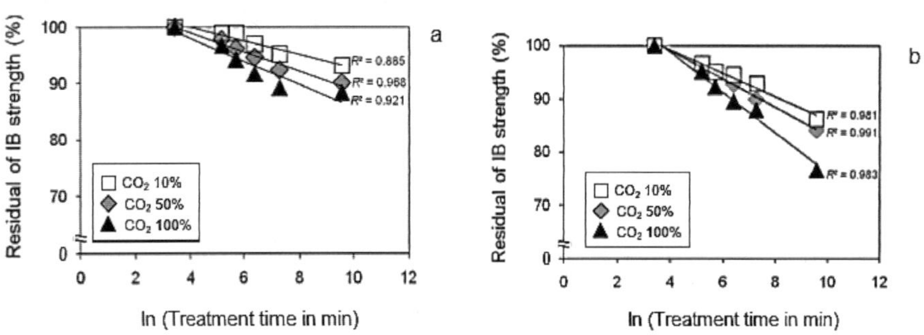


**Figure 1.** Effect of $CO_2$ concentration and pressure conditions on the residual values of internal bond strength (IB) of cement-bonded particleboard (CBP) at various treatment times:
**a** Pressure of 1.0 MPa (Gaseous phase), **b** Pressure of 10 MPa (Supercritical phase)

### *3.2 The residual values of CBP properties*

The residual values of IB strength of conventional cured CBP exposed at different concentration and pressure of $CO_2$ at various treatment times are presented in Fig. 1, which indicates the linear relationships between the residual values of IB strength, and $CO_2$ concentration (%) and $CO_2$ pressure (MPa). As mentioned earlier, the residual values of CBP properties which indicate the degradation of CBP properties from the initial values (100%) was measured based on the degradation values of CBP properties treated at both the gaseous and supercritical $CO_2$ in certain period compared to the optimum values (the initial values) of CBP properties treated in 30 min. Although in our previous research,[9] $\sqrt{t}$ (min) was performed in treatment time, however, in this experiment the treatment time (min) was performed in natural logarithm or *ln* (*x*) to obtain the best coefficient of determination ($R^2$) between the properties of CBP and the affecting factors. Figure 1 shows the residual values of IB strength of CO2-treated boards decreased proportionally when the $CO_2$ concentration and pressure applied were increased over a longer treatment time. At $CO_2$ pressure of 1.0 MPa or in gaseous phase, the residual values of IB strength were obtained 88% at 10 days treated by 100% of $CO_2$ concentration. In contrast, the residual values of IB strength were 93.2% when 10% of $CO_2$ concentration was applied in 10 days

(Fig. 1a). The residual values of IB strength were 90% when 50% of $CO_2$ concentration and 1.0 MPa of $CO_2$ pressure were applied in 10 days. These residual values of IB strength were almost similar to those obtained by conventional CBP treated by 50% of $CO_2$ concentration at 10 MPa in 24 h. A similar trend was observed for conventional CBP treated at 10 MPa of $CO_2$ pressure or in supercritical phase as shown in Fig. 1b. Addition of 10, 50 and 100% of $CO_2$ concentration at 10 MPa of $CO_2$ pressure over the treatment time from 3 h to 10 days resulted in degradation of the IB strength. The residual values of IB strength were 86.3% at CBP treated by10% of $CO_2$ concentration at 10 MPa of $CO_2$ pressure in 10 days. However, the increasing of $CO_2$ concentration to 100% over the treatment time from 3 h to 10 days showed in any decreases of the residual values of IB strength. The result shows the residual values of IB strength were decreased to about 76.6% at CBP treated by100% of $CO_2$ concentration at 10 MPa of $CO_2$ pressure in 10 days.

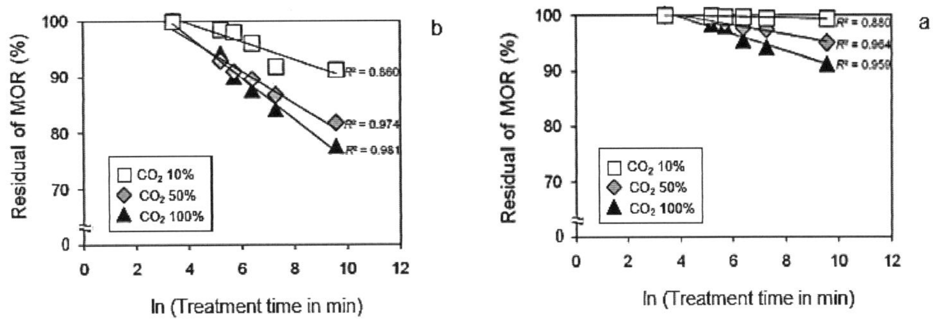


**Figure 2.** Effect of $CO_2$ concentration and pressure conditions on the residual values of modulus of rupture (MOR) of cement-bonded particleboard (CBP) at various treatment times:
**a** Pressure of 1.0 MPa (Gaseous phase), **b** Pressure of 10 MPa (Supercritical phase)

**Table 2.** Multiple regression summary for variables of $CO_2$ concentration and treatment time to CBP properties at different $CO_2$ pressures

| Experiments | Properties | Equation determined | $R^2$ | Corrected $R^2$ |
|---|---|---|---|---|
| 1.0 MPa | IB | IB = 0.0203 ($CO_2$) - 1.64 *ln* (t) + 107 | 0.762 | 0.730 |
| (Gaseous $CO_2$- | MOR | MOR = 0.0042 ($CO_2$) - 2.26 *ln* (t) + 108 | 0.708 | 0.690 |
| phase) | MOE | MOE = 0.0331 ($CO_2$) - 1.57 *ln* (t) + 107 | 0.825 | 0.802 |
| | TS | TS = 0.0458 ($CO_2$) + 1.49 *ln* (t) + 92.2 | 0.805 | 0.779 |
| | WA | WA = 0.0349 ($CO_2$) + 0.974 *ln* (t) + 94.1 | 0.643 | 0.595 |
| 10 MPa | IB | IB = 0.0287 ($CO_2$) - 2.87 *ln* (t) + 109 | 0.893 | 0.879 |
| (Supercritical $CO_2$- | MOR | MOR = 0.0288 ($CO_2$) - 2.86 *ln* (t) + 109 | 0.891 | 0.877 |
| phase) | MOE | MOE = 0.0142 ($CO_2$) - 3.19 *ln* (t) + 113 | 0.903 | 0.890 |
| | TS | TS = 0.0320($CO_2$) + 2.60 *ln* (t) + 88.4 | 0.824 | 0.801 |
| | WA | WA = 0.0381 ($CO_2$) + 1.30 *ln* (t) + 92.4 | 0.713 | 0.675 |

*ln* is natural logarithm, $R^2$ is coefficient of determination

Similar relationship was found in the case of our previous study[9] on the treatment by supercritical $CO_2$ at the longer time span. When 10, 50 and 100% of $CO_2$ concentration were applied at 1.0 MPa of

$CO_2$ pressure over the treatment time from 3 h to 10 days, the degradation of CBP still occurred although the longer treatment time resulted in a small decrease of the IB values. In contrast, when increasing the $CO_2$ pressure to 10 MPa at 100% of $CO_2$ concentration for 10 days, the residual values of IB strength were decreased drastically from 86.3% to about 76.6%. This finding shows that $CO_2$ in supercritical phase degraded the IB strength of CBP in rapid rate with longer treatment time compared to that in gaseous phase.

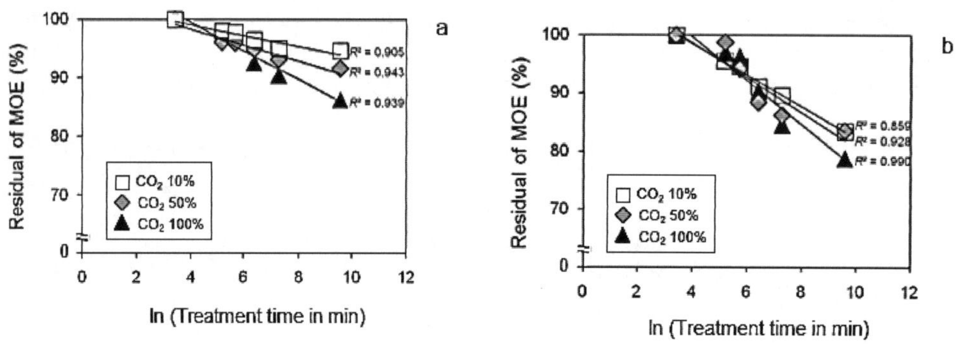


**Figure 3.** Effect of $CO_2$ concentration and pressure conditions on the residual values of modulus of elasticity (MOE) of cement-bonded particleboard (CBP) at various treatment times:
**a** Pressure of 1.0 MPa (Gaseous phase), **b** Pressure of 10 MPa (Supercritical phase)

Figures 2 and 3 show the effect of $CO_2$ concentration and pressure conditions on the residual values of MOR and MOE of conventional cured CBP at various treatment times. The results show the residual values of MOR and MOE were decreased significantly over the treatment time from 3 h to 10 days at 100% of $CO_2$ concentration with 10 MPa of $CO_2$. The residual values of MOR and MOE were 91.7% and 86.2%, respectively, in 10 days, when 100% of $CO_2$ concentration and 1.0 MPa of $CO_2$ pressure were applied (Fig. 2a, 3a). However, the increasing of $CO_2$ pressure to 10 MPa over the treatment time from 3 h to 10 days showed in any decreases of the residual values of MOR and MOE. It was observed that the residual values of MOR and MOE were decreased significantly over the treatment time from 3 h to 10 days at 100% of $CO_2$ concentration with 10 MPa of $CO_2$ pressure. The residual values of MOR and MOE of conventional cured CBP were decreased to about 77.6% and 78.7%, respectively, in 10 days (Fig. 2b, 3b). These results indicated that the degradation of the bending strength of CBP are mainly affected by $CO_2$ concentration, $CO_2$ pressure and treatment time.

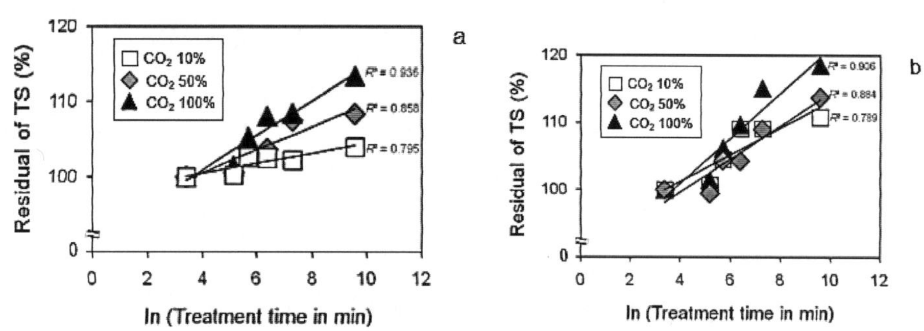


**Figure 4.** Effect of $CO_2$ concentration and pressure conditions on the residual values of thickness swelling (TS) of cement-bonded particleboard (CBP) at various treatment times:
**a** Pressure of 1.0 MPa (Gaseous phase), **b** Pressure of 10 MPa (Supercritical phase)

As clarified in previous researches, [9,10] the presence of carbon dioxide is believed to increase the mechanical properties of CBP during a short treatment time, even after the conventional curing of 28 days,11 however, with a longer treatment time against $CO_2$ in higher concentrations and pressures or in supercritical phase, it was considered that the mechanical properties of CBP degraded rapidly compared to the lower concentrations and pressures of $CO_2$ or in gaseous phase. On the other hand, by exposing CBP to higher concentrations and pressures of $CO_2$, beside increased the mechanical properties of conventional cured CBP in the short treatment time, the degradation process could be accelerated in the longer treatment time over 3 h. The degradation of IB strength of CBP may be due to the degradation on the core-layer of these boards, whereas, the degradation of MOR and MOE of CBP is due to the degradation on the surface-layer. The degradation process was identified due to the ease of penetration and diffusion of supercritical $CO_2$ into the micro pores of the cement matrix caused by the low viscosity and high diffusivity of this media, providing enhanced reaction rate[15] and continuous process of carbonation [16].

Figure 4 shows the effect of $CO_2$ concentration and pressure conditions on the residual values of TS of conventional cured CBP at various treatment times. The residual values of TS of CBP increased with increasing the concentration and pressure conditions. The residual values of TS were 113.4% in 10 days, when 100% of $CO_2$ concentration and 1.0 MPa of $CO_2$ pressure were applied (Fig. 4a ). The residual values of TS of CBP were increased to about 118.4% at 100% of $CO_2$ concentration with 10 MPa of $CO_2$ pressure in 10 days (Fig. 4b). These results show the same condition to WA of CBP, indicating the degradation of dimensional stabilities of conventional cured CBP increased by increasing of $CO_2$ concentrations, $CO_2$ pressures and treatment times. In general, the short treatment time of $CO_2$ in supercritical phase improved the dimensional stability of conventional cured CBP. However, the dimensional stabilities were degraded when these boards treated at 100% of $CO_2$ concentration at 10 MPa of $CO_2$ pressure or in supercritical phase over a longer treatment time, suggesting that the supercritical $CO_2$ treatment over a longer treatment time may result in degradation of dimensional stabilities of conventional cured CBP, which could result from the carbonation process.

### *3.3 The estimation of long term degradation of CBP affected by $CO_2$ in gaseous and supercritical phase*

To estimate long term degradation of CBP affected by $CO_2$ in gaseous and supercritical phases, the following multiple regression model was applied:

$$y = a\ (\mathrm{CO_2\ concentration}) + b\ ln\ (\mathrm{time}) + c \tag{3}$$

where $y$ is the CBP properties (IB strength, MOR, MOE, TS and WS), $a$,$b$ and $c$ are constants, $ln$ is natural logarithm.

The constants $a$, $b$, and $c$ and the coefficient of determination ($R^2$) are summarized in Table 2. For all factors affecting the degradation process, linear relationships were clearly observed. The coefficient of determination ($R^2$) for Eq.(3) between the IB strength of the samples, $CO_2$ concentration and treatment time at 1.0 MPa and 10 MPa of $CO_2$ pressure were 0.762 and 0.893, respectively, indicating that $CO_2$ treatment in both gaseous phase (at 1.0 MPa of $CO_2$ pressure) and supercritical $CO_2$ phase (at 10 MPa of CO2 pressure) affect not only enhanced the mechanical properties of CBP in the short treatment time, but also degraded these properties in the longer treatment time. The $R^2$ values showed a good linearity ($R^2 > 0.891$) for the other properties such as MOR and MOE treated at 10 MPa of $CO_2$ pressure. The $R^2$ values between the TS and WA of the samples, $CO_2$ concentration, and treatment time at 1.0 MPa were 0.805 and 0.643, respectively; and at 10 MPa of $CO_2$ pressure were 0.824 and 0.713, respectively. From these coefficient determination values, it is obvious that the $CO_2$ concentration, $CO_2$ pressure and treatment time affected the degradation of IB strength, MOR, MOE, TS and WA of CBP.

From the multiple regression equation (3), the long term degradation is estimated to obtain the half-value (50%) of initial values (100%) of CBP properties. In gaseous phase or using 1.0 MPa of $CO_2$ pressure, the long term degradation of CBP to obtain half-value of the residual values on the IB strength and MOR were 4.6 × 109 and 2.7 × 105 years, respectively. However, in supercritical phase or using 10 MPa of $CO_2$ pressure, the long term degradation of CBP to obtain the half-value of the residual values on the IB strength and MOR were 3.5 × 103 and 4.7 × 103 years, respectively. These results show the times required for degradation to obtain half-values of such properties treated at 10 MPa of $CO_2$ pressure or under supercritical phase was markedly short compared to the times at 1.0 MPa of $CO_2$ pressure or under gaseous phase. It seemed reasonable to conclude that some degradation occurred at CBP during treatment against $CO_2$ in higher concentrations and pressures or in supercritical phase over a longer treatment time, degrading the mechanical properties and dimensional stabilities of CBP in rapid rate compared to in lower concentrations and pressures of $CO_2$ or in gaseous phase. However, these validities of the above-estimations need further investigation concerning variability of the concentration and pressure of $CO_2$ for exact prediction of long term degradation of CBP, especially at ambient condition.

## 4. Conclusions

The properties of CBP are improved by $CO_2$ treatment in short period in both gaseous and supercritical phases even after the conventional curing process. These properties, however, over the longer treatment time degraded in rapid rate against $CO_2$ treatment in both gaseous and supercritical conditions. It was considered that the degradation of CBP increased with increasing the $CO_2$ concentration, $CO_2$ pressure and treatment time. With the longer treatment time against $CO_2$ in higher concentration and pressure or under supercritical phase, the residual values of mechanical properties of CBP were decreased; indicating CBP performance was degraded rapidly compared to treatment in the lower concentration and pressure of $CO_2$ or under gaseous phase. Although high coefficient determination values between the residual values of CBP properties, and the treatment time and $CO_2$ concentration were observed, and the simple linear model was found to estimate the long term degradation of CBP among the properties and affecting factors, these validities of the estimations need further investigation concerning variability of the concentration and pressure of $CO_2$ for exact prediction of long term degradation of CBP, especially at ambient condition.

**References**

[1] Al-Kadhimi T K H, Banfill P F G, Millard S G, Bungey J H 1996 Accelerated carbonation procedure for studies on concrete. *Adv. Cem. Res.* **47** pp 8

[2] Ngala V T, Page C L 1997 Effects of carbonation on pore structure and diffusional properties of hydrated cement paste Cem. Concr. Res. **27** (7) pp 995–1007

[3] Johannesson B, Utgenannt P 2001 Microstructural changes caused by carbonation of cement mortar *Cem Concr Res* **31** pp 925–931

[4] Arandigoyen M, Bicer-Simsir B, Alvarez Jl, Lange DA (2006) variation of microstructure with carbonation in lime and blended pastes. *Apll Surf Sci* **252** pp 7562–7571

[5] Garcia-Gonzales CA, Hidalgo A, Andrade C, Alonso MC, Fraile J, Lopez-Periago AM, Domingo C (2006) Modification of composition and microstructure of Portland cement pastes as a result of natural and supercritical carbonation procedures. *Ind Eng Chem Res* **45** pp 4985–4992

[6] Garcia-Gonzales CA, Hidalgo A, Fraile J, Lopez-Periago AM, Andrade C, Domingo C (2007) Porosity and water permeability study of supercritical carbonated cement pastes involving mineral additions. *Ind Eng Chem* Res 46 pp 2488–2496

[7] Bertos MF, Simons SJR, Hills CD, Carey PJ (2004) A review of accelerated carbonation technology in the treatment of cement-based materials and sequestration of CO2. *J Hazard Mater* **112** pp193–205

[8] Sisomphon K, Frank L (2007) Carbonation rates of concretes containing high volume of pozzolanic materials. *Cement and Concrete Research* **37** pp 1647–1653

[9] Hermawan D, Hata T, Umemura K, Kawai S, Nagadomi W, Kuroki Y (2001) Rapid production of high-strength cement-bonded particleboard using gaseous or supercritical carbon dioxide. *J Wood Sci* **47** pp 294–300

**Acknowledgments**

The authors thank the Directorate of Higher Education; Ministry of National Education-Indonesia for their finance support by DIKTI scholarship.

IOP Conf. Series: Materials Science and Engineering **593** (2019) 012019 doi:10.1088/1757-899X/593/1/012019

# Colourability of Wood and Its Effect on Bonding Strength of Laminated Wood for Handicraft Material

**N W Dewindiani[1*], Suhasman[1], A D Yunianti[1]**

[1]Forest Products Utilization and Processing Laboratory, Forestry Faculty, Hasanuddin University, Perintis Kemerdekaan Km. 10 St., Makassar, South Sulawesi, Indonesia

[*]Email: nurwidyadewi@gmail.com

**Abstract.** The wood has a high decorative value that can be turned into handicraft products. To optimize the use of small dimensions wood, handicraft products can be made into laminated wood. An attempt to improve the aesthetic value of wood is by adding dyes into the wood. This study aimed to analyse the ability of *Gmelina arborea* (hardwood) and *Pinus merkusii* (softwood) to absorb dyes and the effect of dyeing on bonding strength of laminated wood. Wood planks members (lamellae) that form laminated wood were soaked into direct dyes liquid with soaking time of 0 hours, 1 hour, 6 hours, 12 hours and 24 hours. The results showed retention and penetration dyes of *Pinus merkusii* ranged between 0.00013 g/cm$^3$-0.0003 g/cm$^3$ and 2.98%-8.22%, while retention and penetration of *Gmelina arborea* ranged between 0.00004 g/cm$^3$-0.00018 g/cm$^3$ and 1.59%-4.77%. Retention and penetration tend to increase at dyeing time up to 12 hours. Bonding strength of *Pinus merkusii* ranged between 26.61 kg/cm$^2$-90.05 kg/cm$^2$ while bonding strength of *Gmelina arborea* ranged between 66.97 kg/cm$^2$-96.12 kg/cm$^2$. Based on the result of statistical analysis, type of wood, and the coloration have a real effect on bonding strength of laminated wood. Based on the consideration of resource, efficiency and the results of statistical analysis, 6 hours of dyeing is more efficient in planning of laminated wood material for handicraft products.

## 1. Introduction

Wood as a lignocellulose material can be turned into various purposes, such as raw material of building construction, pulp and paper, composite panel, furniture, handicraft, etc. Handicraft products that utilized small dimension wood are limited to making large size product. Laminated wood is one of developing wood technology in this era because it could utilized small dimension wood that glued together thus becoming large dimension wood. Laminated wood are more efficient because they can be made by combining low-quality wood and high-quality wood [1]

One of the selling points of handicraft product is on its aesthetic value. Wood has a high decorative value that formed from colour, fibre, and pattern on wood [2]. In addition to increase the aesthetic value of wood is by adding dyes into wood to make a different colour of wood natural colour. The effectiveness dyeing will be affected by wood anatomy structure. On the wood, dyes deposited into cavity and micro pore on wood fibre network and suspected can formed hydrogen bonds with hydroxyl group in chemistry component of wood wall cell [3].

Based of this research is to learn the possibility coloration of hardwood and softwood that significantly different in its anatomy. The anatomical structure of wood affected to the ability of a solution penetrate into the wood [4],[5]. Wood that used in this research is *Gmelina arborea* from hardwood group and *Pinus merkusii* from softwood group. Because the wood that has coloured will be processing in different shape, there is a possibility that bonding strength would be affected by dyeing process. Based on that consideration, research about dyeing effect of lamellae on laminated wood bonding strength is deemed necessary.

This research aimed to analyse dyes retention ratio on hardwood and softwood, and analyse the effect of dyeing on bonding strength of laminated wood. The results of this research are expected to be a reference material for laminated wood products innovation. In addition, the results of this research are expected to improve the quality of laminated wood handicrafts products.

## 2. Materials and Methods

Gmelina (*Gmelina arborea*), pine (*Pinus merkusii*), polyvinil acetat (PVAc), epoxy, aquades, and direct dye (Dylon). The tools that used is table saw, pressing machine, digital scales (0.01 g), oven, stereo macroscope, water bath, and universal testing machine.

### *2.1 The preparation of material*

Gmelina and Pine wood was cut and cleaved into 2 × 5 × 20 cm sized. Sized and quality is cultivated same and visually defect free. After that, every field of wood is sanded and dried into oven with 60°C until its water content less than 16%.

### *2.2 Coloring the lamellae*

Axial orientation side of each lamellae are covered with epoxy adhesive to avoid the penetration from longitudinal side. After the epoxy adhesive dried, each lamellae of gmelina and pine are soaked into dyes liquid with concentration of 0.1%. Soaking is done with different time, it is during 0 hours (as control), 1 hour, 6 hours, 12 hours, and 24 hours on water bath with temperature of 100°C. After soaked, each lamellae are dried during 72 hours for each different time of soaked and dried into oven in temperature of 60°C until the water content less than 16%.

### *2.3 Laminated wood making*

Surface of lamellae that already coloured are spread with PVAc adhesive in two sides of bond sides (double glue spread) with 400 g/m$^2$ of glue spread and then assembled. Each laminated wood is formed by two lamellae with same species. After that, laminated wood is pressing during 8 hours under the pressure of 10 kg/cm$^2$.

## 3. Results and Discussions

### *3.1. Retention and penetration of dye*

Retention and penetration of dye on gmelina and pine have different results. Retention and penetration of dye on pine is higher than gmelina, as seen in Figure 1 and Figure 2. This occurs because Pine is more homogeneous in its anatomical structure and has an intercellular canals that allows dye solutions to penetrate into several cell types. But overall retention and penetration in both wood types in this study are very small. This is presumably because the molecular weight of the dye that used is bigger than the pit on the wood and make the dye cannot penetrate into wood. In gmelina, dye is only on the surface of the wood because gmelina does not have an intercellular canals and the pit are very small. The results of Barly and Neo's research of the retention of preservatives in wood showed an increase in solution concentration and the addition of soaking time tends to increase retention with significant differences in the same wood species [6].

In order for dyeing process to take place the molecules of dye should penetrate into the polymer structure. For this (i) the concentration of the dye on fibre surface must be higher than inside the fiber; (ii) the polymer structure should have spaces large enough for the dye molecules to move into fibre interior or, alternatively, the dye molecules should be sufficiently small for this to happened and (iii) molecular chains of fibre polymer should have sufficient freedom of movement in order to facilitate the movement of dye molecules [7]. Low retention and penetration indicate that the second prerequisite is not achieved. The dye polymer is large enough to penetrate the cell wall and move to the next cell wall which is located deeper. Based on GCMS test results of the dye that used, the highest components of dye is $C_{17}H_{34}O_2$ (Hexadecanoic Acid, Methyl Ester) and $C_{19}H_{36}O_2$ (Octadecanoic Acid, Methyl Ester) with molecular weight 270 g/mol and 296 g/mol, these compounds are fatty acid group. This fatty acid is also thought to slow the penetration of dye into wood.

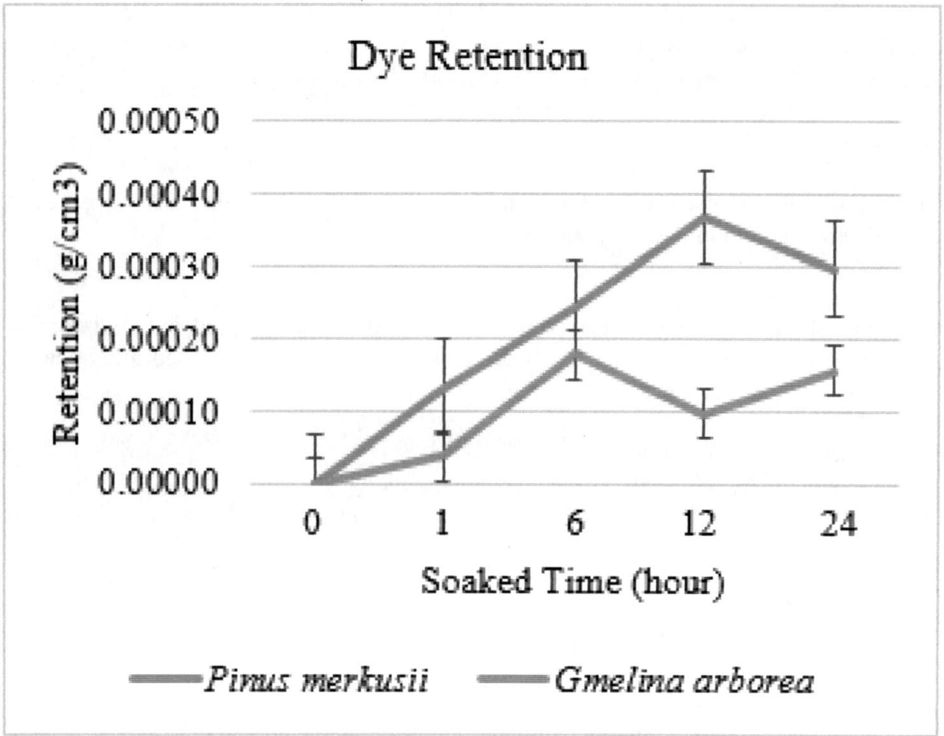


**Figure 1**. The average of dye retention on *Pinus merkusii* and *Gmelian arborea* lamellae

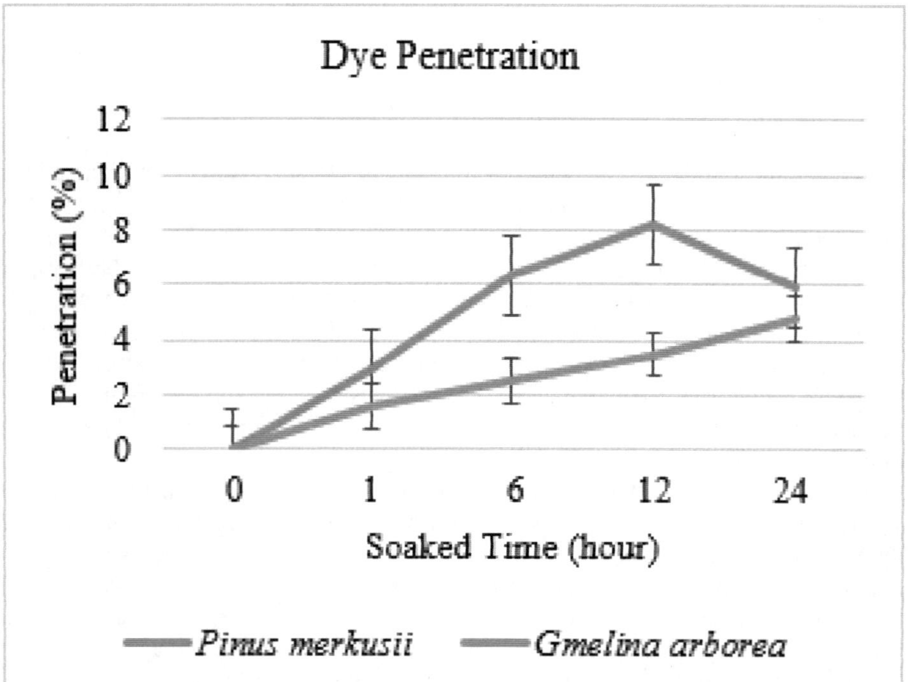


**Figure 2**. The average of dye penetration on *Pinus merkusii* and *Gmelian arborea* lamellae

The data in Figure 3 and Figure 4 show a tendency to increase retention and penetration with increasing soaked time. However the rate of increasing and variability are different. Pine shows the retention and penetration increased from 0 hours to 12 hours of soaked time, then decreased in 24 hours. It could be due to the lamellae at 24 hours of soaked was dominated by heartwood and the radial orientation wood. While the retention of gmelina at 6 hours of soaked is drastic increased. It could be due to the lamellae at 6 hours of soaked was dominated by sapwood and the penetration shows an increase from 0 hours to 24 hours. The wood orientation side (axial, transverse, and longitudinal) of lamellae are vary and the ability to absorb a solutions is also different, because each of these orientation side has a different anatomical structure.

The presence of sapwood and heartwood on the wood also affected to retention and penetration of preservative solution. Heartwood is more difficult to preserve than sapwood, because of its anatomical, physical, and chemical changes that occur during the change of sapwood into heartwood. This change is followed by the death of the sapwood parenchyma cell and the accumulation of resins, tannins and others which provide a permanent colour on the heartwood. This situation make vessel cavity and resin vessel being covered, thus inhibiting the movement of preservatives on wood [8]. Because the concept of the colouring wood is same as preservation of wood, this process is assumed to be same. The differential penetration of a stained solution (for example, 0.5 per cent alcoholic safranin) into the end grain of dried timber allows reasonable detection, the stain being absorbed more readily by the sapwood than the heartwood [9].

To analyse more deeply about how significant the difference of retention and penetration between gmelina and pine, and soaked time, then the variance analysis is carried out. Based on variance analysis results, the retention and penetration of dye is significantly different between wood species. This is because each species of wood has various anatomical structures and physical properties. The anatomical structure of wood that affects retention and penetration is the tracheid, pores (vessels), cells of wood fingers, resin ducts, and wood parenchyma [8]. Based on statistical analyse, 6 hours of soaked time is the best soaked time because the laminated wood would be turned into handicraft product and another further processes (shaved, sanded, etc.), it is expected that the dye penetrates deep into the wood so the colour of dye on wood still could be found.

### *3.2 Bonding strength and damage percentage of laminated wood*

The results of average bonding strength test showed the difference on each soaked time and species wood. As seen in Figure 3, bonding strength of gmelina is more higher than pine for each soaked time. It is because in dyeing process using high temperature and make the resin on pinewood is running out and covered lamellae surface and prevent interlocking process between adhesive and wood surface. There are generally three steps in the process of adhesive bonding, the first is usually preparation of the surface to provide the best interaction of the adhesive with the substrate, the second step is that the adhesive needs to form a molecular-level contact with the substrate surface and the third step is the setting, which involves the solidification and/or curing of the adhesive [10]. In this case, the second step does not occur perfectly.

IOP Conf. Series: Materials Science and Engineering **593** (2019) 012019 doi:10.1088/1757-899X/593/1/012019

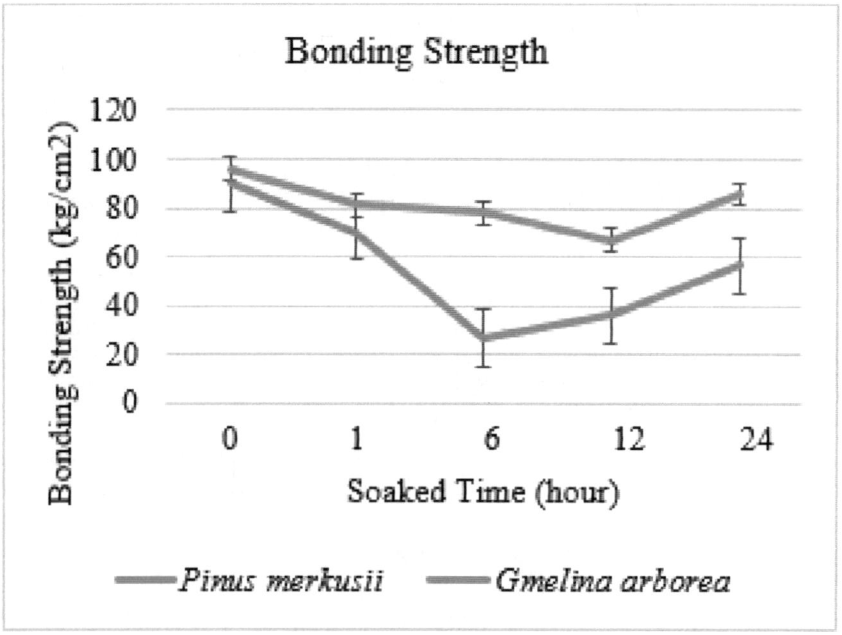


**Figure 3**. The average of bonding strength of *Pinus merkusii* and *Gmelina arborea*

The results of variance analyse showed that soaked time have significant effect on bonding strength of laminated wood. In gmelina, soaking time from 0 hours until 12 hours showed a decrease on bonding strength and at 24 hours of soaking time the bonding strength is increased. While in pine, soaking time from 0 hours until 6 hours showed a decrease on bonding strength and from 12 hours until 24 hours the bonding strength are increased. It is expected caused by dissolution of extractive substances on pine wood after 6 hours of soaked time. Extractive substances is one of the factors that affected on bonding strength, according to research result from Malik and Adi boiling treatment in 35% ethanol solution to extracting extractive substances on Keruing wood could increase the bonding strength of laminated so the results meet the standard requirements for use as indoor components [11]. When wood is treated with inorganic salts, such as boric acid, diammonium phosphate, ammonium sulfate, these chemicals alter the pyrolysis of wood, increasing the amount of char and reducing the amount of volatile, combustible materials. Acidic fi re-retardants can catalyze dehydration of glucose units and depolymerization of cellulose [12]. Pressure treatment with wood preservatives has been known to interfere with the bond integrity of solid wood glued specimens.

The effect of soaking time on bonding strength in this case is more affected by dissolution of extractive substances of wood. Preservative type, preservative retention, and interaction with the surface were reported as highly significant factors affecting shear strength of glue bonds in solid wood samples [13]. Because the method of the colouring method that used is same as preservation method, this system is assumed to be the same. Extractive substance on soaking time of gmelina under 12 hours and pine under 6 hours is expected not reached the wood surface and at the same time the pore is filled with dye thus making the interlocking between adhesive and wood surface. Extractives affected in chemical adhesion by acidity and physics by the condition of the wood surface [14]. Based on JAS 234: 2003 [15], the requirement for bonding strength of laminated wood is at least 54 kg/cm$^2$, pine with soaked time of 6 and 12 hours does not meet the standard.

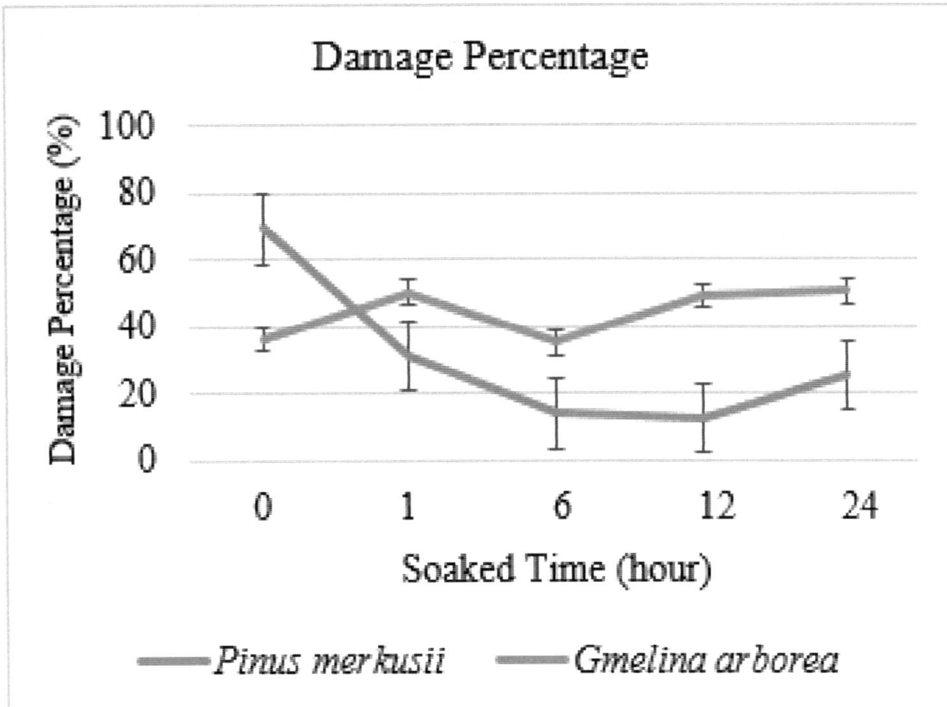


**Figure 4**. The average of wood damage percentage of *Pinus merkusii* and *Gmelina arborea*

The percentage of wood damage in bonding strength test (Figure 4) showed that gmelina has a higher percentage of wood damage than pine in soaking time from 1 hour until 24 hours. While in 0 hour of soaking time (control) gmelina has a lower percentage of wood damage compared to wood. It is because of the resin that covered pine surface prevent the interlocking process between the adhesive and wood surface. The quality of the bonding is stated if the damage mostly occur in wood [16].

## 4. Conclusion

Retention and penetration of dye on pine (softwood) is higher than gmelina (hardwood) for each soaking time. Soaking time of lamellae is affected on bonding strength of laminated wood. Based on the results of statistical analysis and resource efficiency, soaking time for 6 hours is more efficient for material planning of laminated wood as handicraft products.

## References

[1] Abdurrachman, and Hadjib, N 2005 Kekuatan dan Kekakuan Balok Lamina dari Dua Jenis Kayu Kurang Dikenal. *Jurnal Penelitian Hasil Hutan* **23**(2) pp 87–100

[2] Herawati. 2005. Warna alami kayu. https://www.researchgate.net/publication/42320191 [5 Februari 2018].

[3] Sunarto 2008 Teknik pencelupan dan pencapan jilid 2 (Jakarta: Departemen Pendidikan Nasional)

[4] Kusumaningsih K R 2011 Absorption property of preservative on several building woods *Jurnal Wana Tropika* **1**(1)

[5] Febrianto F, C Anne, A Gumilang, F S Yoresta 2014 Distribution of water borne preservative on wood preserved using full cell and empty cell processes *Jurnal Ilmu dan Teknologi Kayu Tropis* **12**(1)

[6] Barly, and Neo E L 2009 Influence of the thickness, solution concentration and soaking times toward the result of timber preservation. *Jurnal Penelitian Hasil Hutan* **28**(1) pp 1–8

[7] Mark M 2011 Handbook of textile and industrial dyeing volume 1: principles, processes, and types of dyes. Woodhead Publishing, United Kingdom.
[8] Supriana N 1973 Masalah spesifikasi pengawetan kayu bangunan. *Berita Hasil Hutan II*, pp 788–793
[9] Bamber R K 1964 Sapwood and heartwood. *Forestry Commission of New South Wales.*
[10] Frihart C R 2005 *Wood Adhesion and Adhesives. In : Handbook of Wood Chemistry and composites* (New York: CRC Press)
[11] Malik J and Adi S 2011 Karakteristik kayu lamina dari kayu keruing berminyak setelah diekstrak. *Jurnal Penelitian Hasil Hutan* **29**(3) pp 271–277
[12] LeVan S and Winandy J E 1990 Effects of fi re-retardant treatments on wood strength: a review. *Wood and Fiber Science* **22**(1) pp 113–131
[13] Humar M, Žlindra D, and Pohleven F 2007 Influence of wood species, treatment method, and biocides concentration on leaching of copper-ethanolamine preservatives. *Building and Environment* **42**(2) pp 578–583
[14] Nawawi D S, Fauzi F, Rustam D, Wasrin S 2005 Bonding properties of some tropical woods in relation to wood acidity. *Jurnal Teknologi Hasil Hutan* **18**(2) pp 47–52
[15] Japanese Agricultural Standard (JAS) 234 2003 *Glued laminated timber* (Japan: Ministry of Agriculture, Forestry and Fisheries)
[16] Herawati E, Massijaya M Y, and Nugroho N 2008 Karakteristik balok laminasi dari kayu mangium (*Acacia mangium Willd.*). *Journal Ilmu Kayu dan Terknologi Hasil Hutan* **1**(1) pp 1–8

IOP Conf. Series: Materials Science and Engineering **593** (2019) 012020 doi:10.1088/1757-899X/593/1/012020

# The Characteristic of Binderless Com-Ply Made from *Paraserianthes falcataria*

**S Suhasman[1*], A D Yunianti[1], S Saad[1]**
[1]Laboratory of Forest Product Utilization and Processing,
Faculty of Forestry, Hasanuddin University, Makassar, Indonesia

*Email : suhasman@yahoo.com

**Abstract**. Oxidation treatment to activate the chemical components of wood to be able to produce bond formation between the particles have been successfully applied in the manufacture of binderless particleboard. Produced particleboard has a high internal bond strength, however, it has a low modulus of rupture. To improve the bending strength, addition of veneer layers on both surfaces are needed. This product is called as com-ply. Because of low accessibility of chemical component of veneer layer compared to particles, then the possibility activation of the chemical components of veneer is still questioned. This study was aimed to investigate the possibility application of oxidation treatment to produce com-ply and to determine the optimum oxidant proportion between particles as the core layer and veneer as surfaces layers. Wood species that used to produce both of particle and veneer were sengon (*Paraserianthes falcataria*). Oxidant levels proportion that used were; 1 : 2, 1 : 3, 1 : 4, 1 : 5, 1 : 6, 1 : 7, and 1 : 0 w/w (veneer : particle). The results showed that the manufacturing of binderless com-ply using oxidation treatment was possible to be applied. Optimum oxidant level proportion between veneer and particle was 1 : 6 w/w. The produced com-ply fulfill the JIS A 5908 2003.

## 1. Introduction

Development of binderless particleboard technology is one of the attractive way to overcome the problems posed by the use of adhesives, in particular formaldehyde emission problem. Some researchers have been intensively developed a variety of methods by utilizing various wood or other woody plant species as raw materials. The methods including; steam injection [1], or enzimatyc activation of particle [2]. In recent years, development of direct hot pressing method using oil palm have been reported by several researchers [3-5]. The other attractive method to produce binderless particleboard using oxidation treatment to activate particle [6]. The produced board has several advantages including high dimensional stability, modulus of elasticity, and internal bond strength. Unfortunately, the modulus of rupture relatively low. This paper is reporting the development of oxidation method in producing binderless panel.

The most common technique to overcome problems related to low modulus of rupture is addition a veneer layer on both of surfaces panel. The products are known as com-ply. The modulus of rupture of com-ply may increases due to the presence of veneer layers. It was occurred, because when the panel receiving bending load, the outer layers will receive highest tensile load. The existence of a thin layer on the outer panel will enhance the boards ability to withstand bending loads significantly. It has been proven that indicated the addition of a layer of thin veneer (20 % of the total thickness of the board) is able to increase the strength of the board almost twice compared to board without veneer layer [6].

This work was focussed on developing high strength and resinless com-ply products using oxidation method. Nonetheless, the feasibility of oxidation method application to produce com-ply is stil questioned. Due to oxidation process need to access the chemical component of wood, so there is no assurance that solid size material such us veneer can be oxidized. This study aimed to evaluate the

IOP Conf. Series: Materials Science and Engineering **593** (2019) 012020 doi:10.1088/1757-899X/593/1/012020

possible application of oxidation treatment in the manufacture of com-ply and determining the optimum oxidant proportion between veneer and particles.

## 2. Materials and Method

Binderless com-ply was produced from air-dried sengon wood (*Paraserianthes falcataria*) particles and veneer. Sengon chips were air dried later converted to a particle with size of 10 mesh passed of sieve using wood mill refiner. Veneer was produced using spindleless rotary veneer. Com-ply was produced by oxidizing the particles and veneer. Before oxidizing, particles and veneer were weighted to produce com-ply with a density target of 0.75 g/cm$^3$ with dimensions of 30 x 30 x 0.7 cm. Particle and veneer then oxidized with hydrogen peroxide 15% based on dry particles and veneer weight and 7.5% of ferrous sulphate based on weight of hydrogen peroxide [6]. Oxidized particle and veneer were conditioned for 15-30 minutes before forming into a mat. The sheets were then hot pressed for 12 minutes at a temperature of 180°C with a pressure of 25 kgf/cm$^2$. The produced com-ply were conditioned for 2 weeks at room temperature, before cutting into samples size according to Japanese Industrial Standard (JIS) A 5908 2003 [7]. Seven types of binderless com-ply were produced in this study based on proportion of oxidant between veneer and particle, namely 1 : 2, 1 : 3, 1 : 4, 1 : 5, 1 : 6, 1 : 7, 0 : 1 (w/w).

## 3. Result and Discussion

The characteristics of com-ply were presented in Figure 1-7. Com-ply density as shown in Figure 1 indicated that only com-ply that produced with the oxidant proportion of 1 : 2 were able to meet the target density of 0.75 g/cm$^2$, while the six other types of com-ply were below the density target. The variability of actual board density is a common phenomenon in the manufacture of composite boards. In some cases, the actual density exceeds the density target while in other cases is lower. Nevertheless, the fact that there is no board that has a density above the target density indicated that the 10% allowance of raw material for producing com-ply is not enough to reach the density target, so the allowance should be increased about 15%. The allowance is intended to anticipate the wasted particles during it handling. Nevertheless, it seems that plenty particles were eliminated when oxidation treatment due to there were particle burned. This is possible occurred due to the exothermic reaction takes place in a high temperature and cause evaporation of water in particle or veneer.

Moisture content of particleboard is presented in Figure 2. It's shows that the overall of com-ply have low moisture content relatively, although its still fullfill the JIS A 5908 [7]. In JIS A 5908 [7], moisture content that allowed of 5-13%. Apparently, oxidation treatment has led to modification of the chemical structure of the raw materials that implicated to hygroscopicity decreases, so the moisture content tend to be lower. In terms of variability of moisture content among of com-ply types, it appears that there is no particular trend that can be attributed to moisture content of com-ply that produced using different variation proportion of oxidant for veneer and particle. This means that differences in the distribution of the oxidant levels in veneer and particle layers did not affect the moisture content.

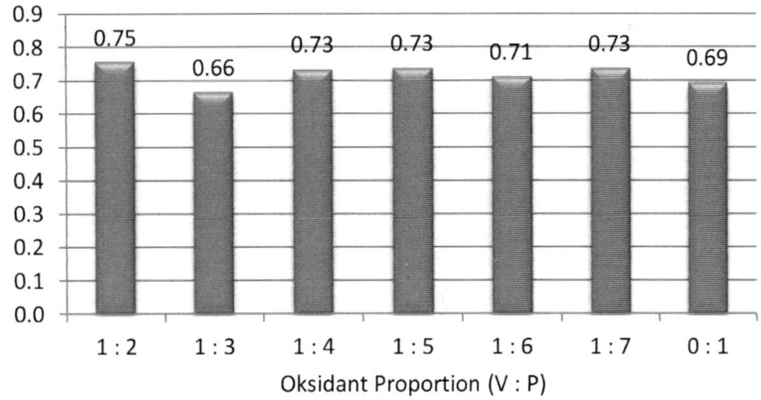


**Figure 1**. Density of Com-ply (g/cm$^2$).

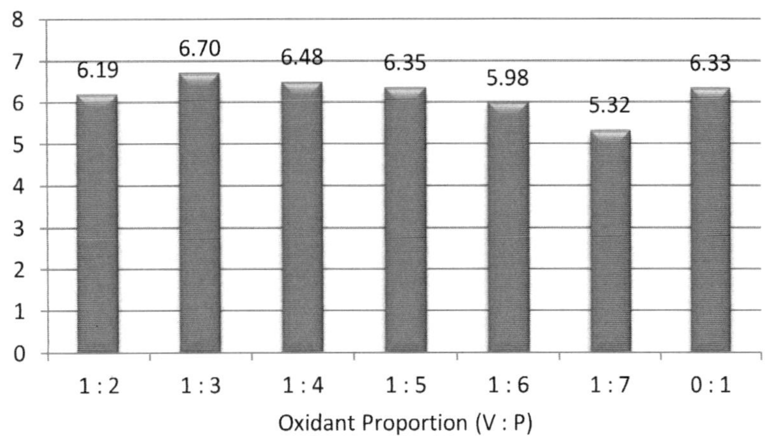


**Figure 2.** Moisture content of Com-ply (%).

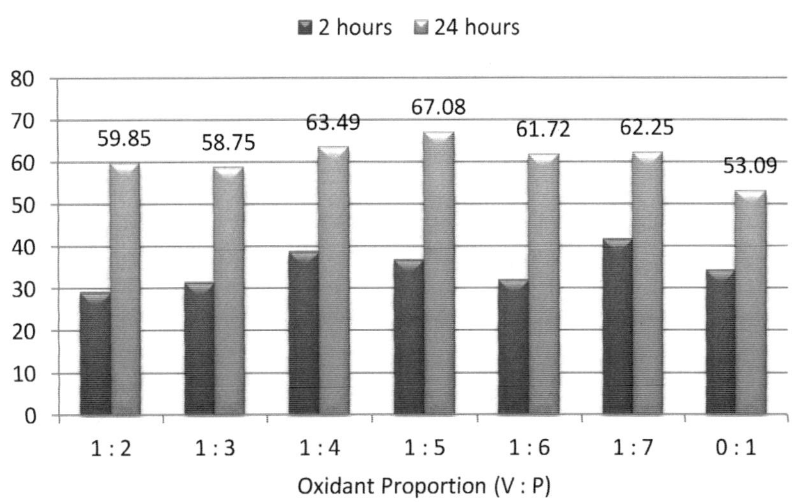


**Figure 3.** Water Absorption of Com-ply (%).

Water absorption of com-ply as presented in Figure 3 shows the similar phenomenon with moisture content. There is no particular trend related to the proportion of oxidant materials in veneer and particle layers. By the time, a lot of oxidant amount relatively in veneer ( 1 : 2 ), water absorption is relatively low and then increased when oxidant proportion in veneer decreases until ratio of 1 : 5. However, after the ratio, water absorption of com-ply tend to decrease again when oxidant in veneer layers increases. This phenomenon illustrates that the level of oxidizing materials in the particles and veneer did not directly related to water absorption.

The data in Figure 3 also shows that the water absorption when the com-ply was immersed for 24 hours indicated that the water absorption range of twice compared with the water immersion for 2 hours. This means that the process of water absorption of com-ply is slow relatively. In general, water absorption within 2 hours of immersion is still much more to fill the empty spaces between the particles then as free water in the cell lumen. In other words, the water takes considerable time to reach the cell wall so that the difference between the water absorption in the immersion for 2 hours and 24 hours is quite large. If the water filling the lumen and cell walls in shorter time, then water absorption differences when immersing for 2 hours and 24 hours may be small. This indicates that the chemical modification due to oxidation treatment has occurred resulting decline hygroscopic materials.

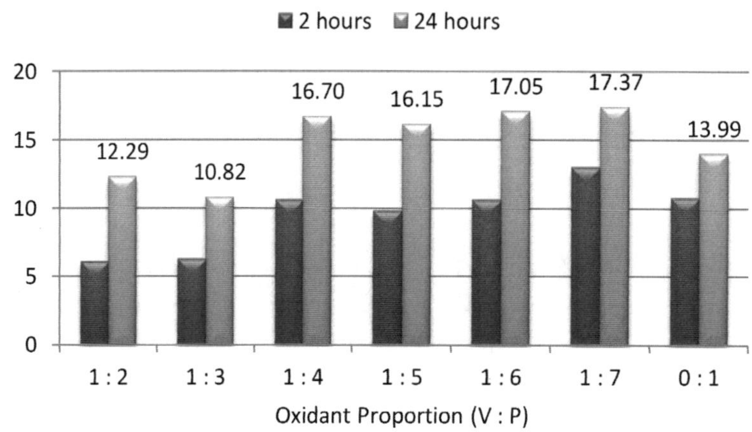


**Figure 4.** Thickness Sweelling of Com-ply (%).

Phenomenon of com-ply thickness swelling similar to that occurring in water absorption, whereas thickness swelling tend to be low when oxidant proportion in veneer is plenty and tend to increase when oxidant in veneer decreases until proportion of 1 : 7. In generally, the relationship between the thickness swelling with water absorption is linear generally. In other words, high water absorption will also result in the high thickness swelling. Based on the data of thickness swelling of produced com-ply, it appears that only com-ply with oxidizing proportion of 1 : 3 between veneer and particles that has a thickness swelling under the maximum limit specified in JIS A 5908 [7] i.e. a maximum of 12 %. This indicates that the dimensional stability of produced com-ply is still quite low. The high thickness swelling seems to be influenced also by the existence of a layer finir. Perhaps, in the process of hot pressing, veneer or particle layers undergo collapse but when immersing in water for 24 hours, so that the cell shave recovered that implicated on high thickness swelling.

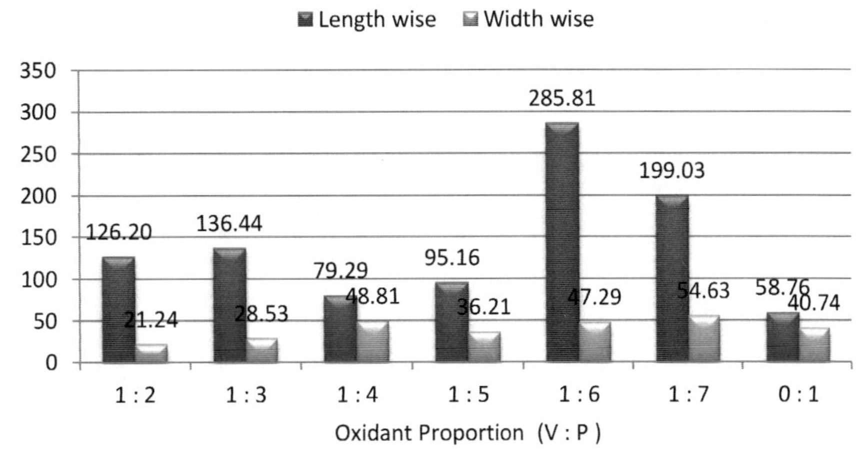


**Figure 5.** MOR of Com-ply (kgf/cm$^2$)

In contrast to the facts that found in the physical properties of boards (density, moisture content, water absorption, and dimensional stability) whereas no certain oxidant proportion that has the best caharacteristict in all physical parameters, in mechanical properties, there is a same tendency in all parameters tested. Com-ply that produced using 1 : 6 oxidant proportion between veneer and particle has the best mechanical properties. The resulting values for all mechanical properties including modulus of rupture, modulus of elasticity and internal bond strength parameters are always highest in that proportion compared to other proportion.

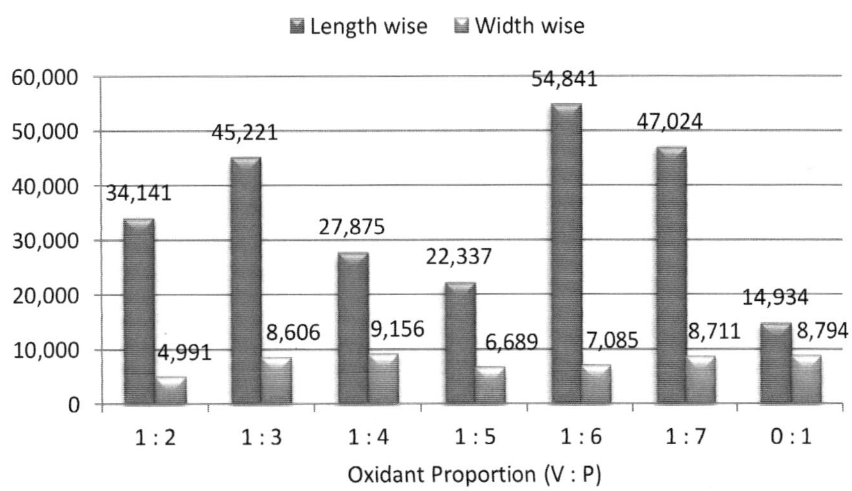


**Figure 6.** MOE of Com-ply (kgf/cm$^2$)

MOR and MOE of the produced com-ply are quite high. The produced sengon binderless particleboard showed that the MOR values that can be achieved just 90 kg/cm$^2$ and MOE of 30,000 kg/cm$^2$ [6]. However, in this study, addition of veneer layer with a thickness of 1 mm for each surfaces to produce com-ply able to produce a high MOR and MOE extremely. The com-ply has three times MOR and almost two times MOE values compared to the binderless particleboard. It is an interesting phenomenon, because then, the addition of a thin layer on the both surface of the board was able to very significant increases of the strength and stiffness of the board.

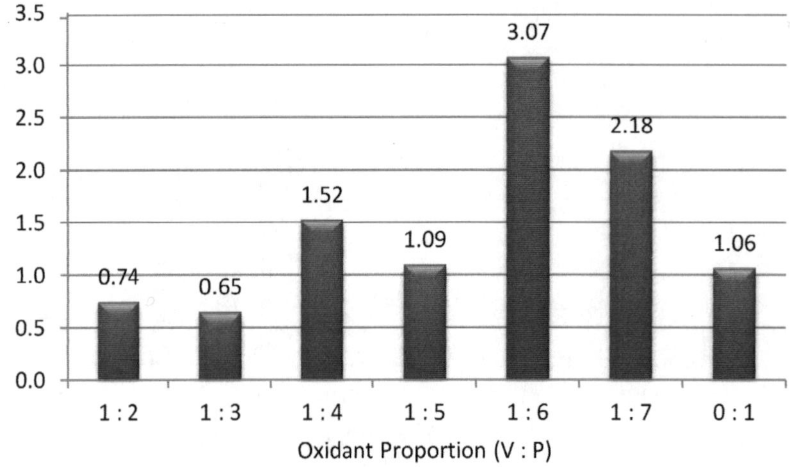


**Figure 7.** Internal Bond of Com-ply (kgf/cm$^2$)

As with MOR and MOE, best oxidant proportion to produce com-ply with high internal bonding strength is the com-ply that produced using oxidant proportion of 1: 6. Based on the mechanical properties data, the com-ply that produced with 1 : 6 proportion meets JIS A 5908 2003 [7]. It can be concluded that in order to produce com-ply, the best oxidant proportion between veneer and particle is 1: 6. However, future effort are still needed to improve dimensional stability of the com-ply. The result of this work indicated that application of oxidation method to produce com-ply is possible to be develoved. By using temperature under 200°C, this method seems quite promising compared with other direct hot pressing methods that developed by using very high temperature. Study about producing binderless particleboard using hot press with temperature more than 200°C [3, 4]. However, similar temperature was used 180°C [5].

## 4. Conclusion

Oxidation treatment to produce binderless com-ply is possible to apply. The best combination of oxidant distribution between veneer and particle is 1 : 6 (w/w). The produced board has high mechanical properties including modulus of rupture, modulus of elasticity and internal bond strength. They fulfill JIS A 5908 2003 [7]. However, the dimensional stability of the com-ply still quite low.

## References

[1] Widyorini R, Higashihara T, Xu J, Watanabe T and Kawai S 2005 Self-bonding characteristics of binderless kenaf core composites Japan *Journal of Wood Science* **39** pp 651–662

[2] Widsten P and Kandelbauer A 2008 Adhesion improvement of lignocellulosic products by enzymatic pre-treatment. Elsevier *Journal of Biotechnology Advances* **26** pp 379–386

[3] Nonaka S, Umemura K and Kawai S 2013 Characterization of bagasse binderless particleboard manufactured in high-temperature range *J Wood Sci* **59** pp 50–56

[4] Saadaoui N, Rouilly A, Fares K, and Rigal L 2013 Characterization of date palm lignocellulosic by-products and self-bonded composite materials obtained thereof *Materials and Design* **50** pp 302–308

[5] Nadhari W N A W, Hashim R, Hiziroglu S, Sulaiman O, Boon J G, Salleh K M, Awalludin M F, Sato M and Sugimoto T 2014 Measurement of some properties of binderless composites manufactured from oil palm trunks and Acacia mangium. *J. Measurement* **50** pp 250–254

[6] Suhasman, Massijaya M Y, Hadi Y S and Santoso A 2011 Particle oxidation time for the manufacture of binderless particleboard. *Wood Res. J.* **2(1)** pp 9–18

[7] Japanese Standards Association (JSA) 2003 Particleboards Japanese Industrial Standard (JIS) A 5908-1994 (Japan: Japanese Standards Association)

IOP Conf. Series: Materials Science and Engineering **593** (2019) 012020 doi:10.1088/1757-899X/593/1/012020

**Aknowledgment**
The author acknowledge to the Directorate General of Higher Education Ministry of Education and Cultural, Indonesia for funding this research under the program of National Strategic competitive research grant (2013-2014).

IOP Conf. Series: Materials Science and Engineering **593** (2019) 012021 doi:10.1088/1757-899X/593/1/012021

# Performance of Particleboard with Varying Jabon Wood and Andong Bamboo Composition Against Subterranean Termite Attack

**W M Arsyad[1*], D R Trisatya, I M Sulastiningsih**

[1]Forest Product Research and Development Center, Ministry of Environment and Forestry, Jl. Gunung Batu No 5, Bogor, Indonesia

[*]Email: waodemuliastuty@gmail.com

**Abstract.** This study investigated the effect of mixed material composition of jabon (*Anthocephalus cadamba* Miq.) and andong (*Giganthochloa pseudoarundinaceae* Steud.) particleboards on their resistance to subterranean termite (*Coptotermes curvignatus* Holmgren). Six types of 35cm × 35cm × 1.2 cm particleboards were produced and glued with phenol formaldehyde in the proportion of 100:0, 70:0, 60:40, 50:50, 40:60 and 30:70 for jabon and andong. Those panels were tested according to SNI 727-2014. The results showed that material composition of jabon and andong was significantly influenced weight loss and termite mortality. Particleboard with 30:70 jabon-andong proportions had the lowest weight loss (4.28%) and classified as resistant class. Termite mortality rate escalated as jabon composition in the panel decreased.

## 1. Introduction

It was reported that forest industries yield a significant amount of waste [1]. This waste could be recycled [1] and offers alternative materials for wood substitution [2]. Wood wastes and other lignocellulosic fibers can be altered into particles and manufactured as composite products [3, 4].

Recent advanced technology on adhesive and manufacturing processes contributed to the increasing demand of composite products for structural and furniture purposes as opposed to solid wood [5]. Further, the advancement technology made it possible to have superior physical and mechanical properties compared to solid wood [5]. Composite products were composite materials which are manufactured by binding lignocellulosic fibers with adhesive [6]. The panel can be manufactured from fibers, particles, strand, veneer, lamina and board of wood [6]. However, those lignocellulosic fibers should met physical, mechanical and biological properties of wood-based products [7]. Particleboards were one of composite products that have been widely utilized ranging from structural wood composites and furniture purposes [8].

Wood-based particleboard commonly had low resistance against wood destroying organisms (9). Extensive studies on treatment for increasing wood and composite products resistance against termite infestation were recorded [10, 11, 12, 13, 7 5]. Nevertheless, research on the resistance of jabon and andong particleboards against subterranean termite attack is lacking. This study was aiming for determining the influence of the proportion of jabon (*Anthocephalus cadamba* Miq.) and andong bamboo (*Gigantochloa pseudoarundinaceae* Steud.) particles on the resistance to subterranean termite *Coptotermes curvignatus* Holmgren attack.

## 2. Materials and Method

### *2.1. Materials*

Jabon and andong bamboo were collected from community plantation in West Java. Particles of jabon and andong were collected from wood working activity and bamboo stripping process. Particles were screened through 5 mm and 2 mm screens. Particles which passed through a 5 mm screen and retained on a 2 mm screen were used. Those particles were then air-dried and oven-dried for 48 hours at 80°C to reach moisture content of 4%. Afterwards, it stored in an air-tight container to keep the moisture content. Liquid phenol formaldehyde (PF) was used as adhesive.

Materials used for laboratory termite resistance test in this study were subterranean termite *Coptotermes curvignathus* Holmgren, sterile sand and distilled water.

### *2.2. Particleboard manufacturing and sample test preparation*

Six variations of particleboard were made with a dimension of 35 cm x 35 cm x 1.2 cm and targeted density of 0.6 g $cm^{-3}$. The panels were manufactured with the following jabon-andong proportions: 100:0, 70:30, 60:40, 50:50, 40:60 and 30:70.

Jabon and andong particles were weighed and mixed in accordance to the assigned proportions. The amount of liquid phenol formaldehyde (PF) adhesive used was 10% (resin solid) on the basis of dry–weight particles. The hardener para formaldehyde was 0.5% of the PF weight and wax emulsion 0.5% of the particles weight was added to PF. A specified quantity of particles was sprayed with certain amount of glue mixture in a blender. The mixture was then hand-formed into a loos mat in a wooden deckle box (30 x 30 cm internal dimensions). The mat then loaded in a hot press set at temperature of 150°C and specific pressure of 25 kg $cm^{-2}$ for 10 minutes. Afterwards, it stored in room temperature for 7 days prior to laboratory test.

### *2.3. Laboratory termite resistance test*

The resistance test was performed in accordance to SNI 7207-2014 [14]. Testing specimens of the particleboards were 2.5 cm × 2.5 cm × 1.2 cm. Five replications were made for each types of particle proportions. The specimens were oven-dried (100±2)°C until a constant weight is obtained and weighed (W1). The specimens were buried in a cylindrical glass tube where 200 g sterile sands which have moisture content below water holding capacity (7%) were placed. About 200 healthy and active workers of subterranean termite were introduced in the cylindrical glass tube and kept in a dark room for four weeks (Figure 1).

**Figure 1.** Subterranean termite test specimen.

After four weeks, the specimens were oven-dried until a constant weight is obtained and weighed. Weight losses and mortality percentage were calculated using Equation 1 and 2.

$$WL = \frac{W_1 - W_2}{W_1} x\ 100 \qquad (1)$$

where: WL= weight loss (%), $W_1$= initial oven-dried weight (g), $W_2$= final oven-dried weight (g)

$$M = \frac{R_n - R_h}{R_n} \times 100 \quad (2)$$

where: M = mortality (%), Rn = number of termite at the initial treatment (subterranean termite = 200 termites), Rh = number of termite after the treatment .

The resistance class of particleboards against termite attack referring to SNI 7207-2014 are presented in Table 1. [14]

**Table 1.** Classification of particleboards against subterranean termites attack (*Coptotermes curvignathus* Holmgren).

| Class | Weight loss interval (%) | Resistance criteria |
|---|---|---|
| I | < 3.5 | Very resistant |
| II | 3.5 – 7.4 | Resistant |
| III | 7.5 – 10.8 | Moderately resistant |
| IV | 10.9 – 18.9 | Non-resistant |
| V | > 18.9 | Susceptible |

*2.4. Data analysis*

Completely Randomized Design was applied to analyze laboratory test results by using SPSS Ver.23. The effect of proportions of jabon and andong particles to the particleboard resistance againts subterranean termite attack were tested, with the following proportions: 100:0 (JA1), 70:30 (JA2), 60:40 (JA3), 50:50 (JA4), 40:60 (JA5) dan 30:70 (JA6). If the proportions of jabon and andong particles significantly affected the resistance of particleboards against termite attack, it then followed by Duncan Test [15].

## 3. Result and Discussion

Analysis of variance revealed that the varying jabon and bamboo particle compositions in the panel have significant effect to weight loss ($F_{calc}$= 8.233) and mortality rate ($F_{calc}$= 179.499). The average of particleboard weight loss and termite mortality rate are presented in Figures 2 and 3.

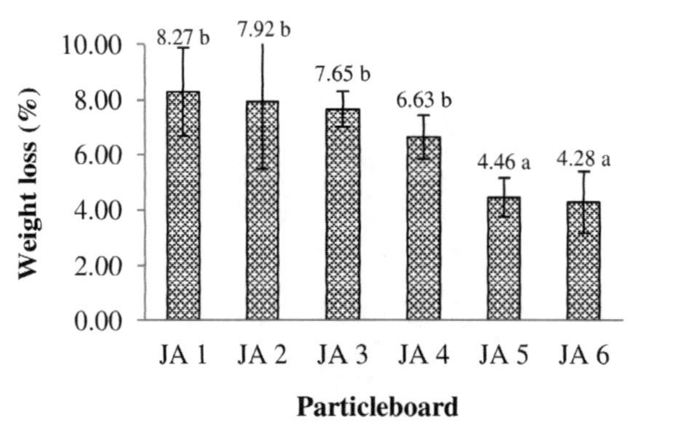


**Figure 2.** Jabon-andong particleboard weight loss.

According to the SNI 7207-2014 on wood testing against wood destroying organism, JA1, JA2 and JA3 were categorized in Class III (moderately resistant), while JA4, JA5 and JA6 were in Class II (resistant) [14]. Figure 2 presents the tendency on increasing weight loss with the addition of bamboo particle. The composition of 70% jabon and 30% andong (JA6) had the lowest weight loss at 4.28%. The highest weight loss was in 100% jabon particleboard (JA1). Hermawan et al. [11] explained that Particleboard composed from high density wood have greater resistance against subterranean termite (*Coptotermes curvignathus* Holmgren) compared to that of low density wood. Wood density of jabon and andong bamboo was 0.34 g $cm^{-3}$ and 0.68 g $cm^{-3}$, consecutively. Particleboard made of more andong particle was more resist to termite infestation than panel with less andong particle. This is in line with previous study which exposed the resistance of mindi wood (*Melia azedarach*) against subterranean and dry-wood termites in comparison to sengon wood (*Paraserianthes falcataria*) due to its distinct density. Wood density was proved has an important role in escalating *Shorea leprosula* particleboard resistance to termite attack [13].

Differences in chemical composition of andong and jabon also affect the resistance of the particleboard. The silica content of andong (1.2%) is higher than jabon (0.1%) [16, 17]. In addition, the extractive content of andong is higher than jabon, this can be seen from the value of solubility in hot water for andong 10.7% while jabon 3.1% [16, 17]. The natural resistance of jabon solid wood was classified as Class V (susceptible) (16), while andong was listed in Class IV (non-resistant) [18]. However, the resistance of a mix jabon and andong particleboard with a certain proportion (30:70) was significantly rose compared to their natural resistance in the solid form. Apart from density factor and its chemical composition, the application of phenol formaldehyde in the panel could augment the resistance class of particleboard against termite attack and reached Class II (resistant). Phenol formaldehyde influenced the resistance of wood-based composite to termite infestation [19, 20 21]. Furthermore, Silva et al. [22] concluded that the chemical content of phenol formaldehyde in *Pinus caribaea* particleboard affected the intensity of termite (*Coptotermes gestroi* Wasmann) infestation.

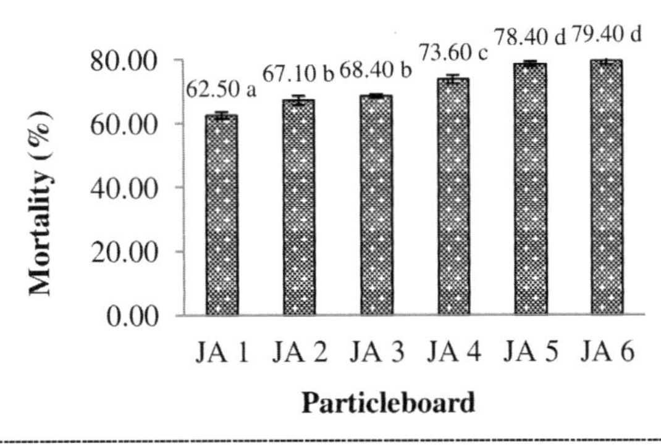


**Figure 3.** Termite mortality rate.

Figure 3 depicts that termite mortality rate increased with the addition of andong particle. The highest mortality rate (79.4%) was in JA6 with 30% jabon and 70% jabon proportion, while the lowest mortality rate (62.50%) was in 100% jabon particle (JA1). Hadikusumo [23] argued that mortality rate over 70% demonstrates wood resistance against termite attack [23]. Furthermore, it was explained that phenol formaldehyde plays an important role in binding less-resistant-wood and bamboo strands. Phenol formaldehyde adhered the strands and its phenol component was harmful to termite [24]. The resin has a distinct odor and strong antiseptic characteristic. Besides poisonous to termite, these characteristics may affect enzyme activities in termite [25]. This may cause the alteration of the feeding activity in termite. The limited available feed resource forced them to kill non-productive termite in the colony [26]. Thus, increases termite mortality.

## 4. Conclusions

Particleboard with 30% jabon and 70% andong yielded the lowest weight loss and the highest termite mortality rate. Thus, this jabon and andong particles proportion was more resistant against subterranean termite attack compared to other particle proportions. This finding suggested that the recycling of jabon and bamboo wastes were potential alternative construction materials due to the growing scarcity of wood supply. The application of renewable resource-based adhesive should also be considered in the further study to produce resistant particleboard.

**References**

[1] Alves S, Silva L da, Mello S A, Azambuja dos A, Varanda M, Donizeti L, Christoforo A L and Rocco F A L 2014 Particleboard produced with sawmill waste of different wood species *Adv. Mat. Res.* **884–85** pp 689–93

[2] Valarelli I D D, Battistelle R A G, Bueno M A P, Bezerra B S, de Campos C I, Alves M C d. S 2014 Physical and mechanical properties of particleboard bamboo waste bonded with urea formaldehyde and castor oil based adhesive *Rev. Mater.* **19**(1) pp 1–6

[3] Silva D A P L, Varanda L D, Christoforo A L, Rocco F A L 2012 Addition of Impregnated Paper Residue to Produce MDP Wood Panel: Example of Solid Waste Recycling International *J. Mater. Eng.* **2** pp 75–9

[4] Santos M F N, Battistelle R A G, Bezerra B S, Varum H S A 2014 Comparative study of the life cycle assesment of particleboards made of residues from sugarcane bagasse (Saccharum spp.) and pine wood shavings (Pinus ellioti) *J. Clean. Prod.* **64** pp 345–55

[5] Ferro F, Santos L, Almeida T, Icimoto F and Lahr F R 2016 Performance of OSB with different preservative treatments to dry wood termites attack *World Conf. on Timber Engineering* Vienna Austria August 22-25 2016

[6] Stark N M, Cai Z and Carll C 2010 *Wood Handbook - Wood as an Engineering Material* chapter 11 pp 1–28

[7] de Melo R R, Stangerlin D M, Campomanes Santana R R, Pedrosa T D 2015 Decay And Termite Resistance Of Particleboard Manufactured From Wood, Bamboo And Rice Husk *Maderas Cienc. y Tecnol.* **17**(1) pp 55–62

[8] Wang S, Yang T, Lin L, Lin C and Tsai M J 2008 Fire retardant treated low formaldehyde emission particleboard made from recycled wood waste *Bioresource Technol.* 99 2072–77

[9] Widyorini R, Yudha A P, Lukmandaru G and Prayitno T A 2015 Sifat fisika mekanika dan ketahanan papan partikel bambu dengan perekat asam sitrat terhadap serangan rayap kayu kering *Jurnal Ilmu Kehutanan* **9**(1) pp 13–22

[10] Köse C, Terzi E, Büyüksari Ü, Avci E, Ayrilmiş N, Kartal S N and Imamura Y 2011 Particleboard and MDF Panels made from a mixture of wood and pinecones: Resistance to decay fungi and termites under laboratory condition *BioResources* **6**(2) pp 2045–54

[11] Hermawan D, Hadi Y S, Fajriani E, Massijaya M Y and Hadjib N 2012 Resistance of Particleboards Made from Fast-Growing Wood Species to Subterranean Termite Attack *Insects* **3** pp 532–77

[12] Acda M N and Cabangon R J 2013 Termite resistance and physico-mechanical properties of particleboard using waste tobacco stalk and wood particles *Int. Biodeter. Biodegr.* **85** pp 354–58.

[13] Indrayani Y, Hardiansyah G and Pari G 2014 Termite resistance of three layer particleboard Shorea leprosula Miq. from natural and plantation forest *Proc. of The 6th Int. Symp. of IWoRS* ed Hartono et al. Medan 12-13 November 2014

[14] Badan Standardisasi Nasional 2014 *Uji ketahanan kayu terhadap organisme perusak kayu SNI 7207–2014* (Jakarta: BSN)

[15] Steel R G and Torrie J H 1995 *Prinsip dan Prosedur Statistika: Suatu Pendekatan Biometrik* (Jakarta: Gramedia Pustaka Utama)

[16]. Martawijaya A, Kartasujana I, Mandang Y I, Prawira S A and Kadir K 2005 *Atlas Kayu Indonesia Jilid II* (Bogor: Media Aksara)

[17] Fatriasari W and Hermiati E 2008 Analisis morfologi serat dan sifat fisik kimia pada enam jenis bambu sebagai bahan baku pulp dan kertas. *Jurnal Ilmu dan Teknologi Hasil Hutan* **1**(2) pp 67–72

[18] Jasni, Damayanti R and Pari R 2017 Ketahanan alami jenis-jenis bambu yang tumbuh di Indonesia terhadap rayap tanah *Jurnal Penelitian Hasil Hutan* **35**(4) pp 289–301

[19] Kartal S N and Frederick G 2003 Decay and termite resistance of medium density fiberboard (MDF) made from different wood species *Int. Biodeter. Biodegr.* **51**(1) pp 29–35

[20] Walther T, Kartal S N, Hwang W J, Umemura K and Kawai S 2007 Strength, decay, and termite resistance of oriented kenaf fiberboards *J. Wood Sci.* **53** pp 481–86

[21] Suhasman S, Hadi Y S, Massijaya M Y and Santoso A 2012 Binderless particleboard resistance to termite attack *Forest Prod. J.* **62**(5) pp 412–15

[22] Silva F C da, Trevisan H, Silva K E D, Lelis R C C and Carvalho A G de 2010 Influence of different adhesive utilized for manufacture of particleboards, under activity by *Coptotermes gestroi* (Wasmann, 1896) (Isoptera: Rhinotermitidae). *Floresta* **40**(2) pp 379–84

[23] Hadikusumo S A 2004 *Pengawetan Kayu* (Yogyakarta: Universitas Gadjah Mada Press)

[24] Febrianto F, Purnamasari I, Arinana, Gumilang A and Kim N H 2013 Steaming effect on natural durability of bamboo oriented strand board against termites and powder post beetle *Jurnal Ilmu dan Teknologi Kayu Tropis* **11**(2) pp 161–69

[25] Sulastiningsih I M and Jasni 2012 The resistance of laminated bamboo boards to subterranean termite (*Coptotermes curvignathus* Holmgren). *Journal of Tropical Wood Science and Technology* **10**(1) pp 23–7

[26] Rismayadi Y and Arinana 2007 *Usir rayap dengan cara baru dan ramah lingkungan* (Jakarta: Gramedia Pustaka Utama)

IOP Conf. Series: Materials Science and Engineering **593** (2019) 012022 doi:10.1088/1757-899X/593/1/012022

# Optimization of SC-$CO_2$ Time Conditions and Permethrin Concentration in Laminated Bamboo Production Using Surface Response Methodology

**Z Uslinawaty[1], M Muin[2], B Putranto[2], S Suhasman[2]**

[a]Departmen of Forestry, Haluoleo University, Jl. H.E.A. Mokodompit No. 1, Kendari 93232 South east Sulawesi, Indonesia;
[b]Department of Forestry, Hasanuddin University, Jl. Perintis Kemerdekaan KM-10, Makassar 90245, South Sulawesi, Indonesia;

**Abstract.** Supercritical carbon dioxide method has starting to be used as environmental friendly method in preserving wood. It is expected to maintain physical and mechanical properties of bamboo. The purpose of this study was to determine the effect of the duration times of supercritical conditions and permethrin preservative concentrations on optimization of shear strength of laminated bamboo impregnated by using supercritical carbon dioxide in the initial process (blades) and final process (laminated bamboo). The RSM-CCD experimental design was used to determine the effect of duration times of supercritical carbon dioxide for 20, 25, 30, 35, 40 minutes and the effect of permethrin concentrations 0.1, 0.3, and 0.5% against the shear strength of laminated bamboos. Response surface analysis, using 10 version of design expert software, has been done to get the optimum condition which would produce maximum of shear strength. The results showed that permethrin concentrations have stronger influenced (based on RSM analysis) on the shear strength and durability of laminated bamboo than the times of supercritical carbon dioxide conditions. Optimization of the two combinations of increasing shear strength and endurance were supercritical conditions = 22.71 minutes, and preservative permethrin concentration = 0.24%. Shear strength of shear strength and decrease in combination weight is 85 kg / cm2 and 0.38%.

## 1. Introduction

Many preservation efforts have been carried out to improve bamboo's durability against powder beetles [1, 2, 3, 4]. The preservation methods that have been carried out have several disadvantages, including: decreased mechanical physical properties of laminated bamboo (mainly shear strength), the amount of preservative used, and produces waste that causes pollution to the environment [5, 6]. Therefore the preservation of environmentally friendly methods need to be developed.

One method of preservation that is environmentally friendly and does not cause dimensional changes in some types of wood, composite wood and laminated bamboo is supercritical carbon dioxide impregnation (SC-CO2) [7, 8, 9]. Preservation of wood and composite wood using the SC-CO2 impregnation method produces unchanging physical and mechanical properties. So preservation with the impregnation method SC-CO2 as a solvent is expected not to reduce the mechanical properties (mainly shear strength) and increase the durability of laminated bamboo. This study sought to optimize the time of supercritical conditions and permethrin preservative concentration to produce quality laminated bamboo.

## 2. Method

### *2.1. Sample Preparation*

Bamboo used is Petung bamboo (*Dendrocalamus asper*) with an age of ± 4 years. Stem diameter is ± 25 cm with a thickness of 1-1.5 cm. The bamboo stem used is that is ± 1m from the base of the stem. 4 stems of bamboo were taken from 4 bamboo clumps from Anduna village, Wolasi sub-district, South Konawe district, Southeast Sulawesi province. Bamboo is cut to a length of 15 cm, width 2.5 cm, and thickness according to the thickness of the bamboo wall. The outer and inner parts are removed. Before gluing, the splits bamboo is air dried at a temperature of 60 ° C for 24 hours.

### *2.2. Laminated Bamboo*

Laminated bamboo is made with the inner outer bonding area. The adhesive used is an isocyanate adhesive prepared according to the technical standards specified by the manufacturer. Before being applied, the two adhesive components, namely resin and hardener were mixed and stirred until blended in a ratio of 100: 15 (based on weight). Before the gluing process, the blade surface is smooth, cleaned of all dirt. All adhesive resurfacing systems are carried out using a spatula, and are applied to both the double spread surface of the lamina with a weight of 250 g / m2. The laminated bamboo is cold pressed / clamped with a 3 hour press time and a pressure of 10 kg / cm$^2$.

### *2.3. Preservation with SC-CO2 and permethrin impregnation*

Preservation with CO2 impregnation uses the Multi-Phase Carbon Dioxide (MP-CO2) [7]. The impregnation treatment of SC-CO2 is applied to the bamboo blade before being laminated. In the production of laminated bamboo. Impregnation begins by inserting the blade in the vessel at 15 ° C and a pressure of about 6 MPa. After the temperature and pressure reach 35 ° C and 8 MPa, the sample is left to stand with various times of supercritical conditions and permethrin concentration. Then the sample was removed from the ship and stored in the desiccator for ± 1 hour. Before testing shear strength and testing resistance to beetle powder, samples were stored for a week.

### *2.4. Experiment Design*

This study uses a central composite design (CCD) with factorial experiments consisting of supercritical conditions and permethrin preservative concentrations. Factor level observed: supercritical condition (A), which consists of three levels, namely 20 minutes, 30 minutes, and 40 minutes and permethrin (B) concentration consisting of three levels of 0.1%; 0.3%; and 0.5%. The level of experiment on each factor (independent variable) is coded to simplify the calculation, with the value of code 1 at the largest level, the value -1 at the smallest level, 0 for the center point, and the axial point (IαI = 1.414). Each factor combination is repeated 2 times, except the center point factor is repeated 3 times.

*2.5. Shear Strength Test*

Making test samples for shear strength testing is carried out at least 7 days after gluing bamboo slats. For each replication made 6 examples of press shear test. Testing of bamboo stickiness for each treatment was carried out according to Japanese Standards [10]. The stickiness test is done by giving the load placed in the parallel direction of the fiber by placing the test sample vertically. The maximum load value is read when the test sample has a 0.5 cm damage to the adhesive line.

*2.6. Weight Loss Test*

In plastic tubs measuring 50 x 25 x 5 cm bamboo is inserted which has been attacked by powder beetles. Then covered with wire sheets measuring 0.5 cm x 0.5 cm. On the wire sheet, sample samples were 19 samples based on the run of the CCD design. Then closed using a mosquito net so that the beetle does not come out. Every week a plastic tub is observed for signs of a powder beetle attack. After 12 weeks the sample was dismantled based on the presence of a powder beetle of 30% in the control sample ie non-treatment sample. The test sample is cleaned of fine powder, then weighed.

## 3. Results and Discusison

The value of shear strength of laminated bamboo impregnated with SC-CO2 in the initial process ranged from 8.15 MPa - 8.87 MPa. The value of shear strength of laminated bamboo is better than some laminated bamboo which has been preserved with certain preservatives [4, 5]. The shear strength of laminated bamboo is different based on the type of bamboo, age, and also the adhesive used.

Bamboo durability can be seen in the value of heavy loss due to the attack of microorganisms. The value of weight loss in this study was very low, ranging from 0% -3.02%. The value of weight loss of laminated bamboo with impregnation of SC-CO2 is equivalent to the weight loss of several types of wood preserved with permethrin and using SC-CO2 as a solvent [4]. The effectiveness of permethrin was very good even at a concentration of 0.04% with a value of weight loss of 0%. Permethrin is a synthetic pyrethroid-based on natural pyrethrins chemistry. This preservative was developed to replace harmful organochlorine insecticides. This pesticide has a very low toxicity in the environment. It was developed Permethrin has been used in wood preservation since 1980 because it is beneficial for the environment, its spectrum of poisoning is extensive, stable in wood and has dual action as repellent and contact insecticides

Based on Annova's analysis, a suitable model for shear strength response in the initial process is a linear model with a p-value of 0.0001 (0.01%), which indicates that the probability of a model error of less than 1% or a linear model has a significant (significant) shear strength response initial process. The lack of fit value of the linear model has a p-value of 0.1610 (16.10%) which shows this model is not significantly different at p> 5%. This value indicates that the linear model is suitable for predicting the effect of supercritical conditions and preservative concentration on shear strength. The model developed is suitable for shear strong responses in the initial process and produces the following equation:

Encoded equation:

$$Y = 8.42 - 0.076\,X_1 - 0.207\,X_2 \quad (1)$$

Actual equation:

$$Y = 8.69 - 0.0076\,A - 1.0348\,B \quad (2)$$

Description: $X_1 = \frac{A - 30}{10}$ $X_2 = \frac{B - 0.3}{0.2}$

A = Time of supercritical condition

B = Permethrin preservative concentration

Based on the model equation it is known that the shear resistance of laminated bamboo impregnated in the initial process will increase as the supercritical condition decreases (A) and decreases the concentration of preservatives (B).

The effect of each factor on the shear resistance of bamboo lamination in the initial process can be seen in the Annova. The factors that influence are the time of encoded supercritical conditions ($X_1$) and the concentration of preservatives ($X_2$). The R-value for the laminated bamboo shear response model that was SC-CO2 pre-impregnated in the initial process was 0.8519. This value means that the effect of variables $X_1$ and $X_2$ on changes in the response variable is 85.19% while the remaining 14.8% is influenced by other variables.

Analysis of variance in weight loss in the quadratic model (order polynomial model 2) obtained a Lack of fit value of 36.54 or $p < 0.001$. This value indicates the model is not suitable to explain the effect of treatment on heavy losses, so it is necessary to revise the model stepwise and produce a value of lack of fit that remains significant. This indicates that there are some components that have not been added. So that from the order 2 polynomial model it is changed to the 3rd order polynomial model by adding a cubic component for the time of encoded supercritical conditions and concentration preserved concentrations of the encapsulated. Then the order 3 polynomial model was revised again to eliminate non-significant components. The choice of order 3 without interaction is the most appropriate model (addition of $A^3$ and $B^3$) so that the value of lack of fit is not significant. The Lack of fit value is 0.71 with p-value 0.51 or $p > 0.05$, indicating that the model does not significantly affect the model, meaning that the model is suitable to explain the effect of treatment on the weight loss of laminated bamboo. So the equation for losing weight in the initial process is as follows:

Encoded equation

$$Y = 0.009 + 0.29\,X_1 - 0.94\,X_2 + 0.23\,X_{12} + 0.75\,X_{22} - 0.30\,X_{13} - 0.059\,X_{23} \quad (3)$$

Description: $X1 = \frac{A - 30}{10}$ $X2 = \frac{B - 0.3}{0.2}$

A = time of supercritical condition

B = Permetrin preservative concentration

Actual equation:

$$Y = 12.69 - 0.92\ A - 18.10\ B + 0.029\ A^2 + 25.59\ B^2 - 0{,}0003\ A^3 - 7.38\ B^3 \quad (4)$$

The most influential factors in losing weight of laminated bamboo after testing with beetles are the length of time encoded supercritical conditions ($X_1$), encoded preservative concentration ($X_2$), quadratic length of the supercritical condition encoded $(X_1)^2$ , quadratic concentration of coded preservative $(X_2)^2$, cubic length of encoded supercritical condition $(X_1)^3$ and cubic encoded preservative concentration $(X_2)^3$. The R-value for the weight loss response is 0.9976, this value means that the effect of variables X1 and X2 on the change in the response variable is 99.76% while the remaining 0.24% is influenced by other variables.

Based on the description above it is known that the shear resistance of laminated bamboo impregnated in the initial process will increase with decreasing the time of encoded supercritical conditions ($X_1$) and decreasing the concentration of encoded preservatives ($X_2$). but weight loss will increase in decreasing the preservative concentration, the cubic time of supercritical conditions and the cubic concentration of preservatives. Both of these are problematic because laminated bamboo which is expected from the results of preservation are those that have a high shear strength and heavy loss from low microorganism attacks. So to get the optimization of the two criteria above, a combination process is carried out. The optimization of the two combinations of increasing shear strength and durability was the time of supercritical conditions = 22.71 minutes, and the concentration of preservative permethrin = 0.24%. The shear strength of the shear strength and weight loss of the combination is 85 kg / cm2 and 0.38%. This condition indicates that the permethrin laminated bamboo preservation using the SC-CO2 impregnation method in the initial process can be carried out only with the time of supercritical conditions 22, 71 minutes and with a concentration of 0.24%.

The relationship between dependent and independent variables is illustrated in the form of a three-dimensional response surface curve constructed based on the analysis of shear strength and weight loss due to the attack of the powder beetle. The three-dimensional graph of the surface response of the shear strength of laminated bamboo is presented in Figure1, and a three-dimensional graph of the surface response to the weight loss of laminated bamboo in the initial process is presented in Figure 2.

Figure 1. explained that when the supercritical conditions of the concentration of preservatives both play a role in reactively decreasing the shear strength. The higher the concentration of preservatives causes the shear strength to decrease. The higher time for supercritical conditions (20-40 minutes) also causes shear strength to decrease (Figure 1).

IOP Conf. Series: Materials Science and Engineering **593** (2019) 012022 doi:10.1088/1757-899X/593/1/012022

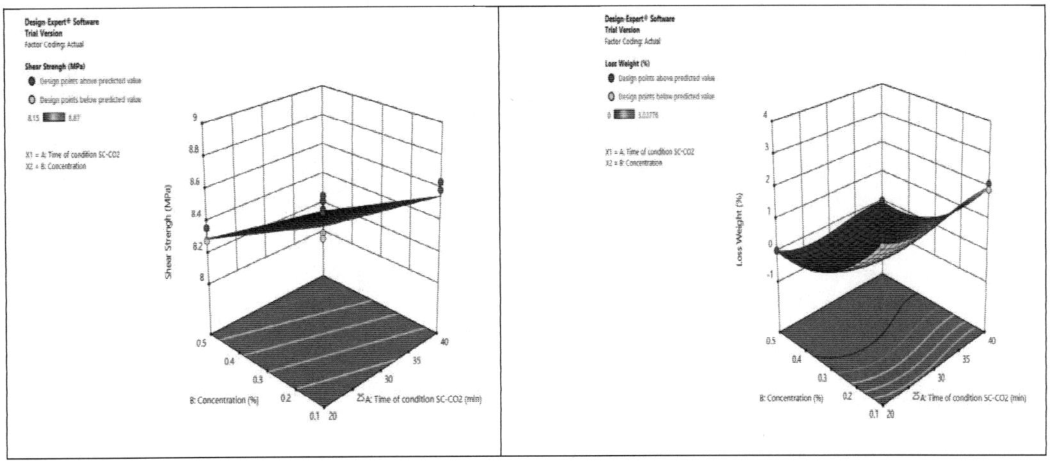


**Figure 1.**

The optimum condition of shear strength was predicted at the time of supercritical conditions of 20 minutes and preservative concentration of 0.1%, with a shear strength of 8.87 MPa. Then after adding time to 40 minutes with a fixed concentration, the shear strength decreases to 8.59 MPa. While the lowest shear strength at the time of 40 minutes of supercritical conditions and 0.5% preservative concentration with a value of 8.15 MPa. This condition shows that the concentration of preservatives plays a role in reactively decreasing the shear strength. Increasing the time of supercritical conditions and increasing concentration, the shear strength will be smaller.

On the three-dimensional curve in Figure 1(a), it will be seen that the concentration of preservatives plays a role reactively reducing weight loss. the higher the concentration of preservatives, the lower the value of weight loss. Conversely, if the preservative concentration decreases, then the weight loss will increase. A good preservation method is a preservation method that is able to increase the resistance of bamboo to the attack of destructive microorganisms and is able to maintain the value of shear strength similar to laminated bamboo which does not undergo a process of preservation.

At the time of 40 minutes supercritical conditions and 0.5% concentration. Bamboo laminates do not experience a heavy loss (0%). When the concentration of 0.5 when the supercritical condition is lowered to 20 minutes, the weight loss occurs at 0.008%. Weight loss increases when the preservative concentration is reduced to 0.1% and the time of supercritical conditions is 20 minutes, weight loss increases to 1.9%. The weight loss value increases again when the supercritical condition is increased to 40 minutes and the concentration is 0.1% to 2.1%.

## 4. Conclusion

- Higher concentration of permethrin preservation, decreased the shear strength but on contrary increased the durability of laminated bamboo.
- RSM used to determine optimal condition of time condition of SC-$CO_2$ and
- The results of this research can be applied to the laminated bamboo industry.

## References

[1] Lee, A. W. C., Chen, G., & Tainter, F. H. 2001. Comparative Treatability Of Moso Bamboo And Southern Pine With CCA Preservative Using A Commercial Schedule. *Bioresource Technology. 77*: 87–88.

[2] Muin, M., and Tsunoda, K. 2003. Preservative Treatment Of Wood-Based Composites With 3-Iodo-2-Propynyl Butylcarbamate Using Supercritical Carbon Dioxide Impregnation. *Journal of Wood Science*. *49*(5): 430–436.

[3] Hanim A.R and anwar UMK, R. 2013. Effects of Chemical Treatments on Durability Properties of Gigantochloa scortechinii Strips and ply-bamboo. *Asian Journal of Biological Sciensce. 6(3):153-160.*

[4] Cookson, L. J., Qader, A., Creffield, J. W., & Scown, D. K. 2009. Treatment Of Timber With Permethrin In Supercritical Carbon Dioxide To Control Termites. *Journal of Supercritical Fluids.* 49 (1): 203-208.

[5] Sulaiman, O., Hashim, R., Wahab, R., Ismail, Z. A., Samsi, H. W., and Mohamed, A. 2006. Evaluation Of Shear Strength Of Oil Treated Laminated Bamboo. *Bioresource Technology. 97*(18): 2466–2469.

[6] Aini, N., Morisco, A. 2009. Pengaruh Pengawetan Terhadap Kekuatan Dan Keawetan Produk Laminasi Bambu. *Forum Teknik sipil. 19*(1): 979–985.

[7] Muin, M. 2011. Physical, Mechanical, and Biological Properties of Silafluofen-treated Tropical Wood Using Multi-Phase Carbon Dioxide Impregnation. *Wood Journal Science. 2*(1): 42–46.

[8] Muin, M., Adachi, A., Inoue, M., Yoshimura, T., and Tsunoda, K. 2003. Feasibility Of Supercritical Carbon Dioxide As A Carrier Solvent For Preservative Treatment Of Wood-Based Composites. *Journal of Wood Science*. *49*(1): 65–72.

[9] Uslinawaty, Z., Muin, M., Putranto, B., and Suhasman, S. 2017. Properties Of Laminated Bamboo Produced With The Involvement Of Supercritical Carbon Dioxide Treatment. *Journal of the Indian Academy of Wood Science*. *14*(1): 74-81. doi: 10.1007/s13196-017-0190-1

[10] Sulastiningsih. 2014. *Pengembangan Papan Laminasi Bersilang Dari Bambu Andong ( Gigantochloa pseudoarundinacea ).* Disertasi tidak diterbitkan. *Bogor.* Institut Pertanian Bogor.

# Medicinal Plant inventory at the Agroforestry Land in Buffer Area of Gunung Leuser National Park

**Rahmawaty[1*], R Amalia[1], R Batubara[1], A Rauf[2]**

[1)]Department of Forest Management, Faculty of Forestry, Universitas Sumatera Utara Jl. Tridharma Ujung No.1, Kampus USU Medan, Sumatera Utara, Indonesia
[2)]Faculty of Agriculture, Universitas Sumatera Utara, Medan, Sumatera Utara, Indonesia

* Email: rahmawaty@usu.ac.id

**Abstract**. The inventory of medicinal plants in North Sumatra Province are important, especially those that are often utilized by the community. This study aimed to map the distribution of medicinal plants (herbs) under the agroforestry land in buffer area Gunung Leuser National Park. This research was conducted in Simpang Kuta Buluh Village, Sei Bingai Sub District, Langkat District, North Sumatra Province. This research was conducted on March to June 2018. The survey method was used by making plots in the field. A Geographic Information System (GIS) was used to analyze the distribution of the medicinal plants and a Global Positioning System (GPS) was used to record the coordinate points of medicinal plants in the area. The results showed that there were found thirty four (34) species of herbs plant in the area with the total of 307 individuals herb,. Sampelulut (Urena lobata L.) was dominant spesies which is found (56 individuals) in this area, followed by Asar-asar (Selaginella deoderleinii) (42 individuals) and Senduduk Bulu (Clidemia hirta) (30 individuals). Constituent plants in Agroforestry land, namely: Mahoni (Swietenia mahagoni), Durian (Durio zibethinus), Karet (Hevea brasiliensis), Pisang (Musa paradisiaca), Angsana (Pterocarpus indicus), Coklat (Theobroma cacao L.), Petai (Parkia speciosa), Cengkeh (Syzygium aromaticum), Asam Glugur (Garcinia atroviridus, and Kelapa (Cocos nucifera). The data of medicinal plants distribution is vital to provide information for community, researchers, and government.

## 1. Introduction

One of the most important biodiversity is medicinal plants. Medicinal plants are currently still widely used by people in traditional medicine and have a very important role for the world of health. Furthermore, medical experts known as physicians make medicinal herbs with raw materials from the forest [3]. The medicinal plants are classified into three, namely: traditional medicinal plants, modern medicinal plants, and potential medicinal plants [13, 14].

One place that has the potential to be found in Simpang Kuta Buluh Village, Sei Bingai Sub District, Langkat District, North Sumatra Province. Simpang Kuta Buluh village is one of the villages in the TNGL area buffer and has good natural medicinal plant potential [2]. Community life still uses plants for traditional treatment. Most of the Simpang Kuta Buluh people have used medicinal plants for generations. There are several medicinal herbs that have long been known by the people of Simpang Kuta Buluh Village. In this research, the community used plants for Traditional medicinal plants, which are plant species that are known or believed to have medicinal properties and have been used as raw materials for traditional medicine, but until now information about the potential of medicinal plants in Simpang Kuta Buluh village is limited, so it is necessary to identify the medicinal plants in the village.

Through this research, it is expected to provide information about medicinal plants that have the potential to be developed commercially, for example as raw materials for certain drugs. Research on medicinal plants in this location has never been done. Based on the description above, this research needs to be done to determine the types of medicinal plants found under the stands in Simpang Kuta Buluh Village, Sei Bingai Sub District, Langkat District, North Sumatra Province and map the distribution of medicinal plants. This study aimed to map the distribution of medicinal plants (*herbs*) under the agroforestry land at Buffer aera Gunung Leuser National Park in Simpang Kuta Buluh Village, Sei Bingai Sub District, Langkat District, North Sumatra Province.

## 2. Materials and Methods

This research was conducted in Simpang Kuta Buluh Village, Sei Bingai Sub District, Langkat, North Sumatra, Indonesia (Figure 1). This research was conducted on March to June 2018. The survey method was conducted to collect medicinal plants samples in the field. The Geographic Information System (SIG) was used to show the spatial distribution of medicinal plants and a Global Positioning System (GPS) was used to record the coordinate points of medicinal plants. The map represents the distribution of medicinal plants in Simpang Kuta Buluh village. To retrieve plant coordinate points by taking one coordinate point that represents all the same medicinal plants that were in the observation plot.

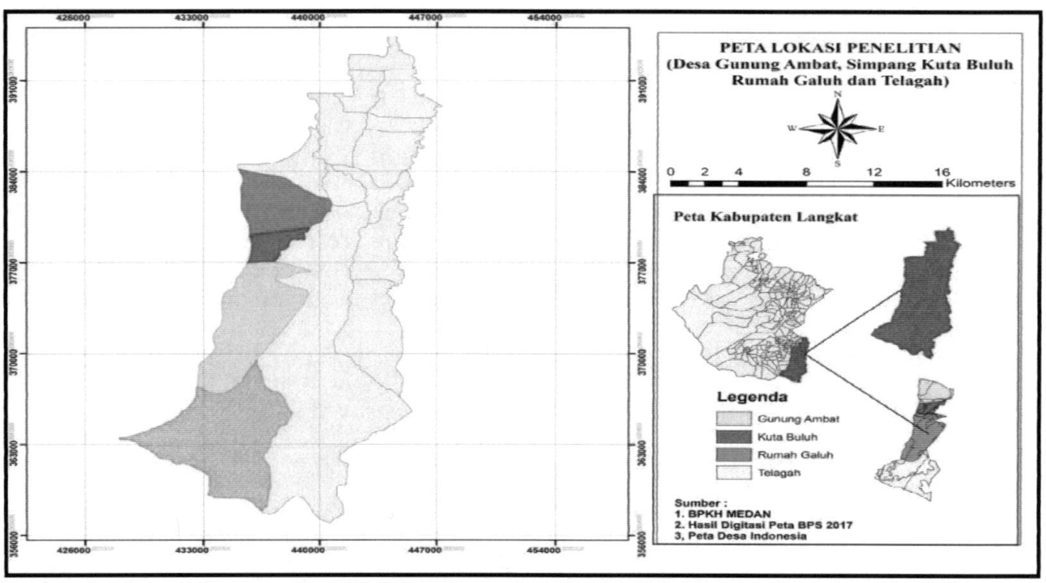


**Figure 1.** Map of Simpang Kuta Buluh Village, Sei Bingai Sub District, Langkat, North Sumatra, Indonesia.

The inventory method of medicinal plants is carried out using a plot sampling method. Determination of the starting point of the inventory in the pathway is done by purposive sampling method, then carried out by systematic sampling, to find out how the state of distribution of medicinal plants in the research area. Collection of medicinal plants using the plot sampling method, namely by making plots in the pathway with sampling intensity of 10% of the land area. Each path is made with a plot of 20 x 20 meters [6]. The total number of plot samples made was 14 plot.

The composition of tree species was determined by the important value index (IVI). The IVI was the sum of the relative frequency (RF), relative density (Rde) and the relative dominance (Rdo). Mastery level of the species in the community indicated by the IVI [6]. The diversity of a forest area can be described by the Shannon Index [6].

## 3. Results and discussion

IOP Conf. Series: Materials Science and Engineering **593** (2019) 012023 doi:10.1088/1757-899X/593/1/012023

Species numbers of medicinal plants, important value index (IVI), and the Shannon Index (H') were presented in Table 1.

**Table 1.** Species numbers of Medicinal Plants, important value index (IVI) and H'.

| No | Local Name and Scientific | Total | IVI | H' |
|---|---|---|---|---|
| 1 | Asar-asar (*Selaginella deoderleinii*) | 42 | 21.7 | 0.27 |
| 2 | Bancir (*Bidens pilosa*) | 5 | 5.63 | 0.07 |
| 3 | Balik Angin (*Mallotus paniculatus*) | 1 | 1.66 | 0.02 |
| 4 | Bas-bas (*Coleus aemboinicus*) | 4 | 2.64 | 0.06 |
| 5 | Belasih (*Lantana camara*). | 7 | 4.94 | 0.09 |
| 6 | Betak (*Paederia scandes* L. Merr) | 21 | 10.8 | 0.18 |
| 7 | Burley (*Zingiber pupureum* R.) | 7 | 3.61 | 0.09 |
| 8 | Bibi kacimukmuk (*Leucas lavandulifolia* ) | 1 | 1.66 | 0.02 |
| 9 | Dulpak (*Blumea balsamifera*) | 1 | 1.66 | 0.02 |
| 10 | Empu-empu (*Lilium* sp.) | 1 | 1.66 | 0.02 |
| 11 | Kacibini (*Elephantopus scaber*) | 10 | 7.25 | 0.11 |
| 12 | Karpe Balu (*Trema orientalis*) | 5 | 4.29 | 0.07 |
| 13 | Kemusing (*Portulaca oleracea* L.) | 5 | 2.96 | 0.07 |
| 14 | Kuning gajah (*Maranta arundinacea*) | 6 | 4.62 | 0.08 |
| 15 | Kurmak-kurmak (*Sterculia javanica*) | 3 | 3.64 | 0.05 |
| 16 | Merah mata (*Hedyotis corimbosa* L.) | 2 | 3.32 | 0.03 |
| 17 | Nungke (*Eurycoma longifolia*) | 2 | 1.98 | 0.03 |
| 18 | Panggil-Panggil (*Peperomia pellucid* L.) | 2 | 1.98 | 0.03 |
| 19 | Pakis (*Polystichum setiferum*) | 5 | 4.29 | 0.07 |
| 20 | Pesul (*Piper betle* L.) | 2 | 3.32 | 0.03 |
| 21 | Pupuk Mulajadi (*Cylea barbata*) | 2 | 1.98 | 0.03 |
| 22 | Ratah Flower (*Eleusine indica*) | 2 | 3.32 | 0.03 |
| 23 | Rimbatasik (*Rauwolfia verticillata*) | 13 | 10.9 | 0.13 |
| 24 | Salantam (*Graptophyllum pictum*) | 14 | 5.89 | 0.14 |
| 25 | Sambung nyawa (*Gynura procumbens*) | 2 | 1.98 | 0.03 |
| 26 | Sampelulut (*Urena 3obate* L.) | 56 | 30.2 | 0.31 |
| 27 | Senduduk Bulu (*Clidemia hirta*) | 30 | 19.1 | 0.23 |
| 28 | Serai (*Cymbopogon citratus*) | 22 | 9.83 | 0.19 |
| 29 | Sikerbeng (*Centella asiatica* L. Urban) | 13 | 8.23 | 0.13 |
| 30 | Suntil-suntil (*Achyranthes aspera*) | 5 | 2.96 | 0.07 |

| | | | | |
|---|---|---|---|---|
| 31 | Solur tangga (*Acalypha australis* Linn) | 3 | 2.31 | 0.05 |
| 32 | Tanduk reFruit (*Phyllanthus niruri* L.) | 2 | 1.98 | 0.03 |
| 33 | Tegi-tegi (*Euphorbia hirta* L.) | 5 | 4.29 | 0.07 |
| 34 | Temu-temu (*Boesenbergia rotunda*) | 6 | 3.29 | 0.08 |
| | Total | 307 | 200 | 2.91 |

Various types of tree were presented in Table 2 and Figure 2.

**Table 2**. Species numbers of Trees.

| No | Lokal Name | Scientific Name | Total |
|---|---|---|---|
| 1 | Angsana | *Pterocarpus indicus* | 2 |
| 2 | Asam Glugur | *Garcinia atroviridus* | 1 |
| 3 | Coklat | *Theobroma cacao L.* | 6 |
| 4 | Cengkeh | *Syzygium aromaticum* | 1 |
| 5 | Durian | *Durio zibethinus* | 10 |
| 6 | Karet | *Hevea brasiliensis* | 103 |
| 7 | Kelapa | *Cocos nucifera* | 1 |
| 8 | Mahoni | *Swietenia mahagoni* | 12 |
| 9 | Pete | *Parkia speciosa* | 1 |
| 10 | Pisang | *Musa paradisiaca* | 17 |
| | Total | | 76 |

Species numbers and important value index (IVI) of medicinal plants at the Simpang Kuta Buluh Village, Sei Bingai Sub District, Langkat District, North Sumatra Province was Presented in Table 1. Based on Table 1, Sampelulut (*Urena lobata L.)* has the highest IVI (30.23 %), followed by Asar-asar *(Selaginella deoderleinii)* (21.7 %) and Senduduk Bulu (*Clidemia hirta) (19.1%).* The value of the diversity index for the study sites was 2.91 (classified as moderate). This means that environmental conditions were still relatively stable. Sampelulut (*Urena lobata L.) was effective for the treatment of itching, fever, dysentery, diarrhea and rheumatism,* Asar-asar *(Selaginella deoderleinii)* was efficacious to reduce heat, improve blood flow, antitoxic, antineoplasm, stop bleeding and eliminate swelling, and Senduduk Bulu (*Clidemia hirta) was* efficacious for treating sugar, vaginal discharge, diarrhea, mouth sores and burns [4].

The advantages of medicinal plants are raw materials found in nature and do not require large costs in their use. The treatment using herbs with traditional medicinal herbs has advantages in general is considered safer than the use of modern medicine [1]. This is because traditional medicine has relatively fewer side effects than modern medicine. Based on WHO, IUCN and WWF records, more than 20,000 species of medicinal plants are used by 80% of the population worldwide. Research reports and literature no less than 2039 species of medicinal plants originating from Indonesian forests [14].

The constituent stands of agroforestry land consist of: Mahoni (*Swietenia mahagoni)*, Durian *(Durio zibethinus)*, Karet *(Hevea brasiliensis)*, Pisang *(Musa paradisiaca)*, Angsana *(Pterocarpus indicus)*, Coklat *(Theobroma cacao* L.*)*, Petai *(Parkia speciosa)*, Cengkeh *(Syzygium aromaticum)* dan Asam Glugur *(Garcinia atroviridus)*and Kelapa *(Cocos nucifera).*

### *3.1 Mapping of Medicinal plants Distribution*

Mapping of the distribution of medicinal plants is carried out by using the method of inventory and retrieval of medicinal plants in the field. Data on the distribution of medicinal plants are stored in GPS in the form of a waypoint, then overlaid with a map of the location of the research, namely the village

IOP Conf. Series: Materials Science and Engineering **593** (2019) 012023 doi:10.1088/1757-899X/593/1/012023

of Simpang Kuta Buluh, Sei Bingai sub-district. Mapping of medicinal plants is done to determine the location of the distribution of medicinal plants in the village of Simpang Kuta Buluh. The aim is to provide information on the potential of medicinal plants in the field and present them in the form of maps to the public. So that when conducting a field survey, people find it easier to find medicinal plants in the Buluh village of Simpang Kuta. Map of the distribution of medicinal plants can be seen in Figure 2-7.

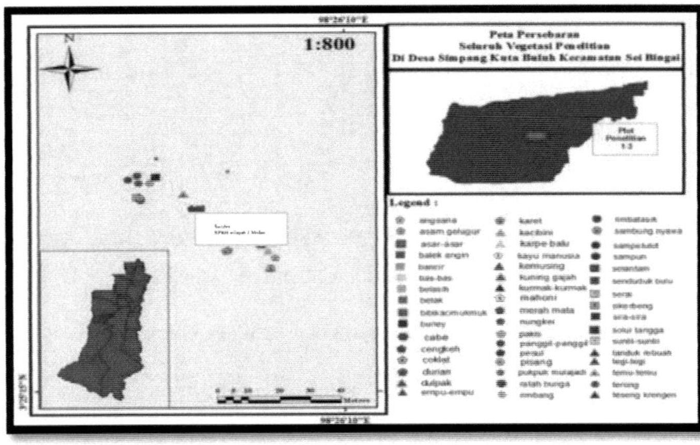


**Figure 2.** Map of Vegetation Distribution in Plot 1, 2, 3.

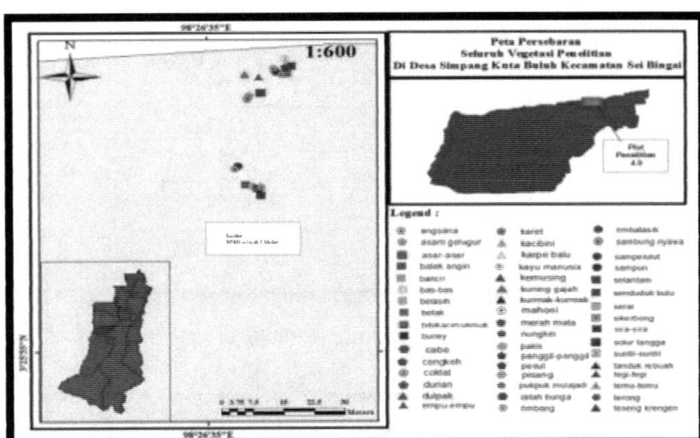


**Figure 3**. Map of Vegetation Distribution in Plot 4 – 9.

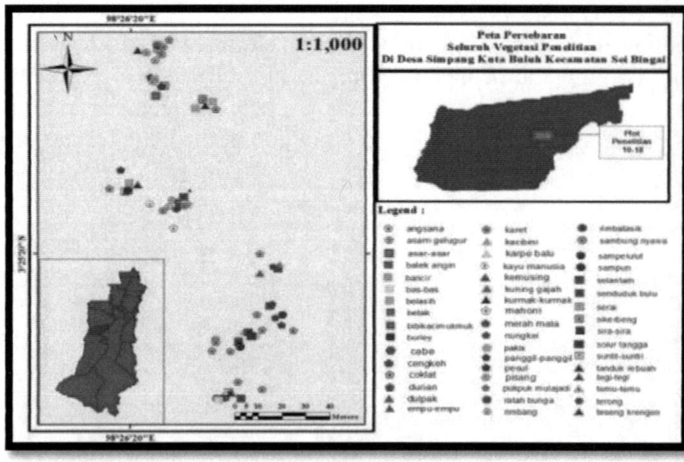


**Figure 4**. Map of Vegetation Distribution in Plot 10 – 18.

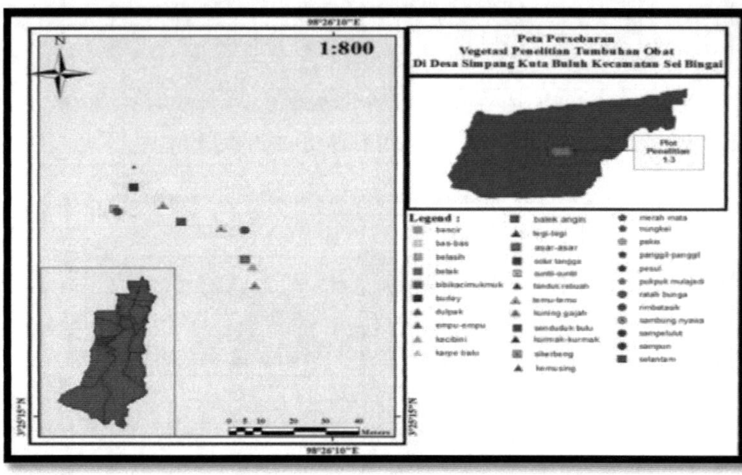


**Figure 5**. Map of Medicinal plants Distribution in Plot 1, 2, 3.

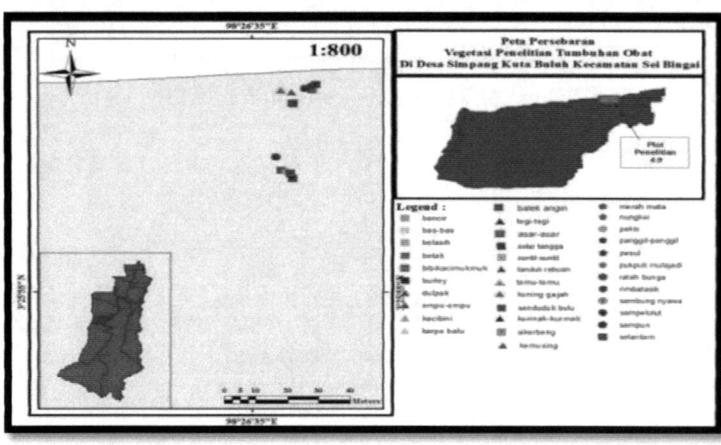


**Figure 6**. Map of Medicinal plants Distribution in Plot 4 – 9.

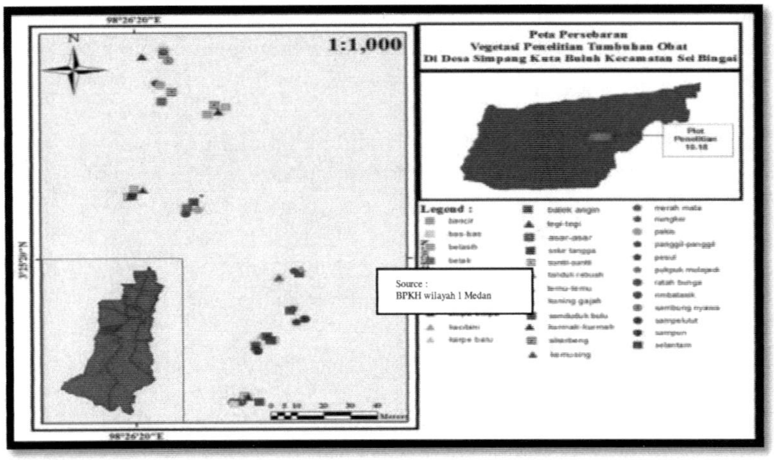


**Figure 7.** Map of Medicinal plants Distribution in Plot 10 – 18.

In Simpang Kuta Buluh Village, there were various types of vegetation have been found that have the potential to become medicinal plants. A total of 307 medicinal plant points were recorded the coordinates points using GPS. It was found that medicinal plants in the village of Simpang Kuta Buluh had a species diversity index of 2.91. The diversity of medicinal plants was classified as moderate ($1 \leq H' \leq 3$), meaning that the ecology of understorey vegetation was in a relatively stable condition [5].

## 4. Conclusion

1. There were found thirty four (34) species of herbs plant in the plot area of Simpang Kuta Buluh Village with the total of 307 individuals herb,. Sampelulut (*Urena lobata* L.) was dominant spesies which is found (56 individuals) in this area, followed by Asar-asar (*Selaginella deoderleinii*) (42 individuals) and Senduduk Bulu (*Clidemia hirta*) (30 individuals).
2. There were found ten (10) species of trees in the plot area of Simpang Kuta Buluh Village with the total of 76 individuals tree, namely: Mahoni (*Swietenia mahagoni*), Durian *(Durio zibethinus)*, Karet *(Hevea brasiliensis)*, Pisang *(Musa paradisiaca)*, Angsana *(Pterocarpus indicus)*, Coklat *(Theobroma cacao* L.*)*, Petai *(Parkia speciosa)*, Cengkeh *(Syzygium aromaticum)* dan Asam Glugur *(Garcinia atroviridus)*.
3. The diversity of medicinal plants in the plot area of Simpang Kuta Buluh Village was classified as moderate.

## Acknowledgments

We are grateful to the Universitas Sumatera Utara for providing financial support (Through USU "Talenta" funding), as well as to the organizing committee of the the Fourteenth Pacific RIM BIO-BASED Composite Symposium 2018, for supporting the presentation of this work on 29-31 October 2018, at the Swiss Bell Hotel, Makassar. We also thank the students of the Department of Forest Management, Faculty of Forestry USU (especially for Tike, Jon, Eral, Novi, Deby, Maden, Uya) who assisted with field work and collect the data.

## References

[1] Abdiyani S 2008 Keanekaragaman Jenis Tumbuhan Bawah Berkhasiat Obat di Dataran Tinggi Dieng *Peneltian Hutan dan Konservasi Alam* **6** pp 79–92.
[2] Djauhari E.and Herani 2004 *Gulma Berkhasiat Obat* (Jakarta: Swadaya)
[3] Kusmana C 2017 *Survey methods and vegetation data interpretation* (Bogor: IPB Press) p 68
[4] Rahmawaty T R, Villanueva, and M G Carandang 2011 *Participatory Land Use Allocation, Case Study in Besitang Watershed, Langkat, North Sumatera, Indonesia.* (Jerman: Lambert

Academic Publishing) p 200

[5] Rahmawaty, E K Sari, A Syofyan, and A Rauf 2014 Integrated Geographic Information System and Global Positioning System for Mapping of Forest Plants in Supporting Natural Resources Protection

[6] Rahmawaty, N C Siregar, and A Rauf 2016 Land Suitability For Tectona Grandis; "Case Study In Arboretum Kwala Bekala, Sumatera Utara University (Kesesuaian Lahan Tanaman Jati; "Studi Kasus Di Arboretum Kwala Bekala, Universitas Sumatera Utara) *J Penelitian Ekosistem Dipterocarpa* **2**(2) pp 73–82.

[7] Rahmawaty, N A Sitorus, and A Rauf 2017a Distribution, above-ground biomass and carbon stock of the vegetation in Taman Beringin Urban Forest, Medan City, North Sumatra, Indonesia *Malaysian Forester* **80**(1) 2018

[8] Rahmawaty, P Patana, and S Latifah 2017b Spatial analysis on distribution of green belt to reduce impacts of climate change in Medan City, North Sumatra. Malays *Appl. Biol.* **46**(2) pp 67–76

[9] Rahmawaty, I E P Sembiring, R Batubara, and P Patana 2018 *Mapping of Tree Damage Classification in the Western Part of Medan City Green Belts Using Geographic Information System* IOP Conference Series: Earth and Environmental Science iopscience.iop.org/volume/1755-1315/166

[10] Zuhud E A M 2009 Potensi Hutan Tropika Indonesia Sebagai Penyangga Bahan Obat Alam untuk Kesehatan Bangsa. Accessed from : http:// www. researchgate. net./ publication/267857735. [22 August 2018] [19.00 WIB].

IOP Conf. Series: Materials Science and Engineering **593** (2019) 012024 doi:10.1088/1757-899X/593/1/012024

# Improving Degradation Ability of Composite Starch/Chitosan by Additional Pineapple Leaf Microfibers for Food Packaging Applications

**I Mutmainna[1], D Tahir[1]*, P L Gareso[1], S Ilyas[1], A Saludung[2]**

[1]Department of Physics, Hasanuddin University, Makassar 90245 Indonesia
[2]Department of Civil and Environmental Engineering, Hiroshima University
1-4-1 Kagamiyama, Hiroshima, Japan

*Email: dtahir@fmipa.unhas.ac.id

**Abstract.** Structural, mechanical properties, and bonding characteristic of plastic bio-composites (starch/chitosan) with an additional Pineapple leaf microfibers (PLM) for 3%, 6% and 9% were carried out by using X-Ray diffraction (XRD), tensile strength test, and Fourier transform infra-red (FTIR) spectroscopy, respectively. The value of tensile strength shows highest for plastic bi-composite with 9% PLM and also XRD spectra shows the amorphous phase with new structure was formed for all bio-composite + PLM. FTIR spectra shows the strong bonding formation as the effect of PLM indicated by increase the intensity of C-H and C=C bond with increasing the amount of PLM. From the biodegradation test shows the PLM strongly influenced to the degradation ability of bio-composite and the highest is for 9% PLM (>80% ). In addition, the lowest browning index is about 37.5% for bio-composite with additional 9% PLM which was determined by using fresh cherry (*Muntingia calabura. L*). It's indicated that the plastic bio-composite with an additional PLM shows high potentials for new food packaging applications.

## 1. Introduction

Environmental problems are increase in every years and become a great concern, due to the increasing number of problems in every year. Environmental effect is the primary reason for the research and development of functional products based on natural resources [1]. Bioplastics are plastics developed from natural resources such as biomass materials from agricultural which were environmentally friendly [2].

Previous studies reported bioplastics based natural resources to form composite such as; PVA / chitosan [3], PVA / fiber pineapple [4-5], and gelatin / chitosan [6]. The focus of their research was about mechanical properties, biodegradation, the level of transparency, effect of temperature and other properties [7-8]. Based on these studies, the alternative base materials for bioplastics are starch, which is low prices, and easily decompose [9]. The disadvantages of bioplastics based on the starch are low mechanical properties, which can be overcome by the additional of chitosan and microfiber from the nature resources as reported in Ref. [10-11]. In Ref. [12] was reported, the bioplastic with addition (1% and 2%) chitosan, shows good mechanical properties [12].

In this study we have used starch and chitosan to form composites and additional Pineapple leaf microfiber (PLM) to increase the mechanical properties, strong bonding formation, and stable structural properties. These properties strongly influenced to the degradation ability, which was determined by using Texture Analyzer MCT - 2150, FTIR spectroscopy, and XRD spectroscopy, respectively. In addition we determined also browning index and biodegradable test for measure the level of moisture and the biodegradation ability by microorganism, respectively.

## 2. Experimental Program

### *2.1 Materials*

Glycerin ($C_3H_8O_3$) (molecular weight 92.09, boiling point $290^oC$) (Merck), acetic acid ($CH_3COOH$) with concentration of 6% (Merck), chitosan (molecular weight 200 Kda, particle size 200-300 mesh, viscosity 55.31 MPa.s, deacetylation rate 94.88%), corn starch (Merck), and aquades (Merck). The acetic acid was diluted from 6% to 1%.

### *2.2 Preparation of pineapple leaf microfiber*

Pineapple leaf microfibers (PLM) are obtained from pineapple leaves by cutting in size of 15-20 cm and heated for 2 hours at $100^oC$ until the color was changed and the leaves were cooled down to room temperature. The PLM were obtained by slicing the cooked pineapple leaves using a cutter. The microfibers were soaked with sodium hydroxide (NaOH) solution 1 Molar for 60 minutes and cleaned using distilled water, and then heated again at a temperature $70^oC$ for 90 minute, the PLM ready for further use.

### 2.3 Bio-composite Sample Preparations

For characterization by XRD, FTIR spectroscopy, and biodegradation test, the prepared ratio bio-composite between starch and chitosan is 50% / 50% with additional PLM 3%, 6%, and 9%,. We prepare also for ratio of the bio-composites are 0/100, 45/55, 50/50, 55/45, 65/35 for comparison and for characterization of tensile strength. We have added the first solution by mixing the starch with glycerin and distilled water for 30 minutes stirring at 95°C and called starch solution, similar for the second solution but use chitosan mixed with 1% acetic acid, chitosan solution. Both solutions were mixed by using magnetic stirrer for 10 minutes to form composite. The PLM with different amount (3%, 6%, and 9%) was added into the composite solution. Finally, composite + PLM solution was formed and then heated at 120° C for 10 hours. For clearly illustration of synthesis process in this study is shown in Figure 1.

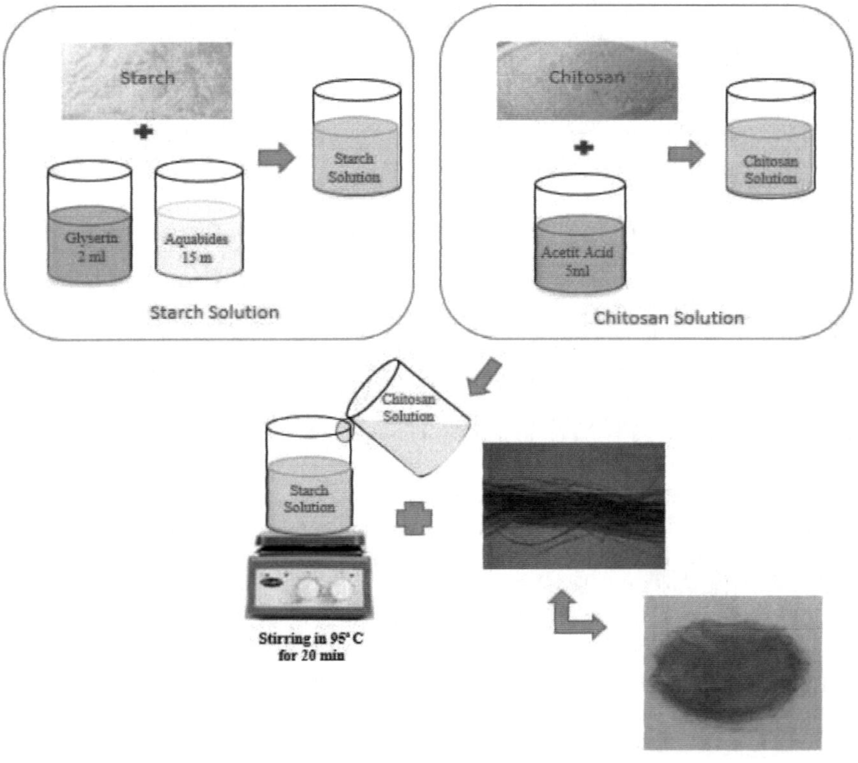


**Figure 1.** Synthesis processes of the composite with additional pineapple leaf microfiber (PLM).

*2.4 Characterization*

The X-Ray diffraction (XRD) was used for analyzing the structural properties, (Shimadzu 7000 X-ray diffraction) with CuKα radiation (λ = 1.5405 Å) for 2θ from 15 ° to 60 °, operates at 30 kV and 10 mA. The time for composite bioplastics to be decomposed was observed by stockpiling the sample under the ground for 3 days, 7 days, 14 days, and 30 days. All samples were weighted for initial weight of the samples and then weighted again after stockpiling the samples to determine the final weight. The bonding characteristics of the sample are determined by the FTIR spectroscopy and mechanical properties were investigated through tensile strength test. The degree of discoloration is determined by using testing the browning index of the fresh cherry (*Muntingia calabura.L)*. We have divided the fresh cherry into 4 groups with three groups are packaging with composite with different composition of starch / chitosan with 9% of PLM and one group used as a control. The weight and the index of browning of cherry are calculated after the 7th days packaged, this method similar reported in Ref. [3]. The color index was determined by following the quantity: 1 for no brownish, 2 for less than 1/4 brownish, 3 for less than ¼ - 1/2 brownish, and 4 for more than 1/2 brownish. Browning index is calculated using the following formula:

$$\text{Browning Index} = \sum \frac{(browning\ scale\ x\ Ni)}{N} \quad \text{...........................} \ (1)$$

where Ni and N are the number of fruits at the level of maturity and the total number of fruits in the treatment, respectively [3].

## 3. Results and discussion

*3.1 X-Ray Diffraction (XRD)*

Figure 2 shows the XRD spectra of starch, chitosan, and bio-composite for starch/chitosan ratio is 50/50 with PLM 3%, 6%, and 9%.

Starch contain two components, first is amylose which has a linear chain structure and the second is amylopectin which has a branch chain structure [8]. The branch structure is the main contribution to form amorphous structures [13-16]. Chitosan also has a linear polymer chain structure, which is usually contributed to forms a crystalline phase by the connection each chains to form regularly polymer [16-17]. In Figure 2, diffractogram (A) shows the structure of starch material which is usually characteristics peaks of amorphous structure at 15° to 25°. Difractogram (B) shows chitosan material which shows more broad peaks at 25° to 35° and small sharp peak indicating the crystalline phase at 48° to 55°. The diffractogram C, D, and E are the XRD spectra for bio-composite (starch/chitosan: 50/50) with the addition of 3%, 6%, and 9% of PLM, respectively, which is similar with diffractogram (B) indicated strong effect of chitosan in the bio-composites.

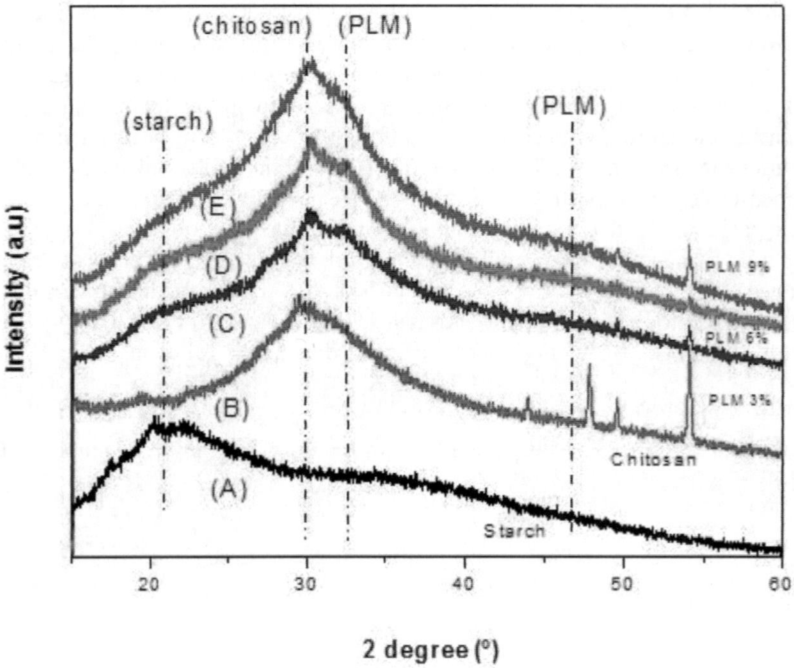


**Figure 2**. X-ray Diffraction (XRD) spectra of starch (A), chitosan (B), bio-composite (Starch/Chitosan 50/50) with additional pineapple leaf microfiber (PLM) 3% (C), 6% (D), and 9% (E) in this study.

The main peak at 30° for bio-composite (starch/chitosan + PLM) is contributed from the starch and peak at around 20° from chitosan, diffraction peak at 34° is contributed from combination of chitosan and PLM and the broad peak at around 48° contributed only from PLM. From these diffraction peaks for composite (starch/chitosan + PLM) indicated the new structure was formed as the contribution from the each base materials.

*3.2 Tensile Strength*

Figure 3 shows the tensile strength for plastic bio-composites with additional PLM; 3%, 6% and 9%.

Figure 3 shows the additional of pineapple leaf microfiber was improved the tensile strength with highest values is 51,659 MPa for ratio composite starch/chitosan is 65/35 with PLM 9%. This result confirms that the pineapple leaf microfiber (PLM) play important role in increasing the tensile strength of composite bioplastics.

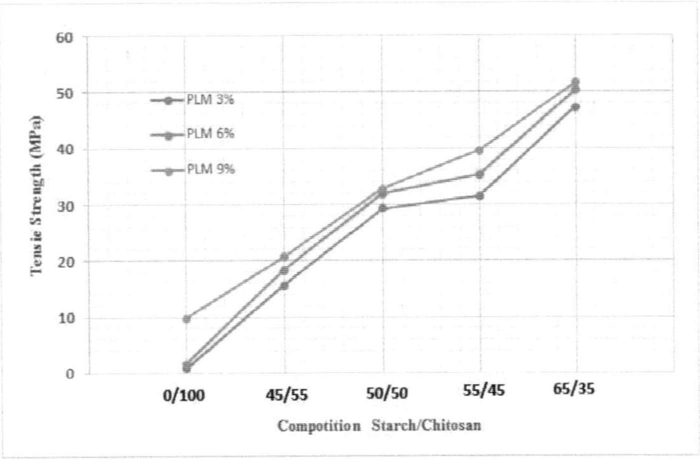


**Figure 3**. Tensile strength of composite bio-plastic in this study

*3.3 FTIR analysis*

FTIR characterization was carried out to determine the bonding characteristics of the composites bioplastic with additional PLM 3%, 6%, and 9% for weight ratio of composite starch/chitosan is 50/50.

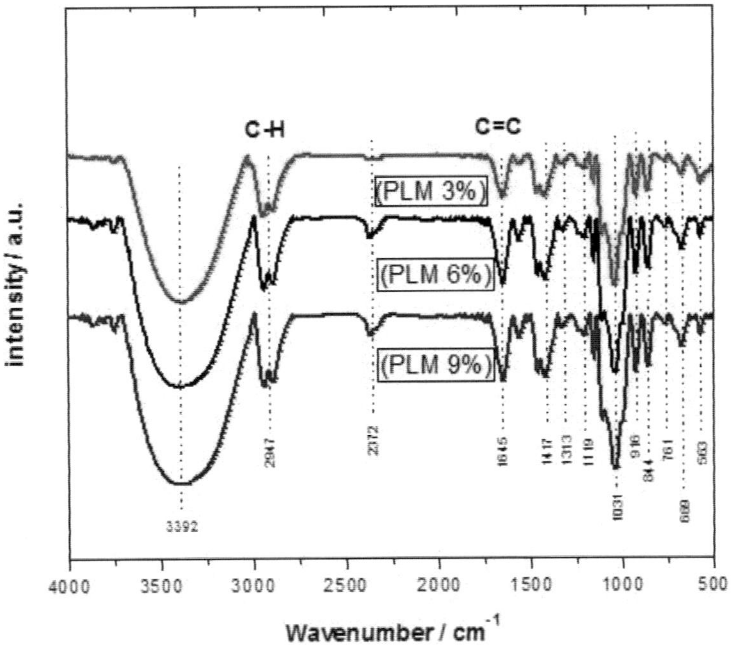


**Figure 4.** Fourier transform infra-red (FT-IR) spectra for composite bioplastic (Starch/Chitosan) is 50/50 with additional PLM (3%, 6%, and 9%).

The FTIR spectrum from the previous study shows that the bonding formation is contribution from the constituent raw material [18]. Figure 4 shows the FTIR spectra with broad absorption peak at 3392 cm$^{-1}$ indicated that the presence of hydroxyl groups in the bio-composites. The C-H bonding at wave number 2917 cm$^{-1}$, and C=C bonding at 1645 cm$^{-1}$ which is show the effect of PLM, and similar for C-H bonding at 1417 cm$^{-1}$. C-O bonding was formed at 1051 cm$^{-1}$, and another. C-H bonding were formed at 916 cm$^{-1}$,

IOP Conf. Series: Materials Science and Engineering **593** (2019) 012024 doi:10.1088/1757-899X/593/1/012024

811 $cm^{-1}$, and 761 $cm^{-1}$, respectively. The absorption intensity of C=C and C-H bonding was increases with increasing the amount of PLM in composites indicated the strong bonding formation was formed. The C=C, and C-H, bonding may strong relation in increasing the mechanical strength and the degradation of bioplastic as reported in reference [19] for polymeric materials.

*3.4 Biodegradation Test*

The results of biodegradation test of bio-composite with additional PLM 3%, 6%, and 9% for 3, 7, 14 and 28 days as can be seen in Figure 5. The degradation ability as a function of time from 3 days to 28 days is clearly effective for high amount of PLM. Composite with 9% PLM show the higher percentage of degradation >80% for 28 days may due to the increasing of C=C and C-H bonding formation in plastics bio-composites. These results confirm that the addition of PLM is significantly affected to the decomposition process. Bioplastics degraded in the environment for the very short period of time probably due to the present of C = C and C-H bond as the effect of PLM which was also contributed to the mechanical properties and the browning index.

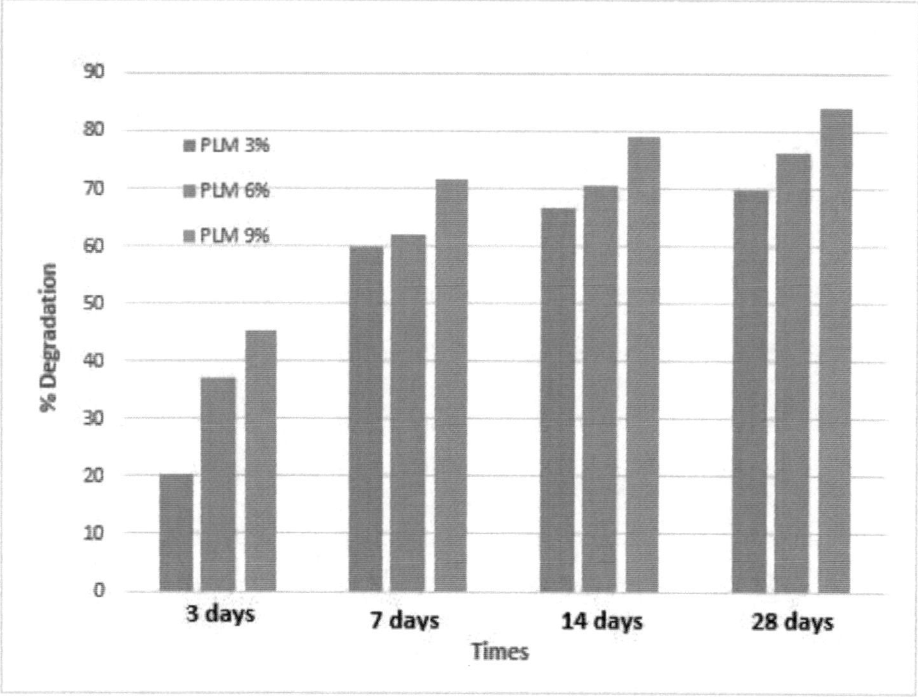


**Figure 5**. Biodegradation test for composite bioplastic (Starch /Chitosan: 50/50) with additional PLM (3%, 6%, and 9%)

*3.5 Measurement Browning Index*

We have used ratio of composite is 50/50% + 9% PLM and also for ratio 50/50, 55/45 and 65/35 for comparison in determining the browning index in this study. The decreasing weight and the discoloration of the cherry (*Muntingia calabura.L*) fruit were observed after 7 days of packaging. For comparison, we used the data control from fruit without packaging.

**Table 1**. Browning index of composite bioplastic for composition 50/50, we have included compositions 55/45 and 65/35 with 9% PLM for comparison.

| Bio-composite + PLM 9% | Weight Loss (%) | Browning Index (%) |
|---|---|---|
| **Control** | 0.540 ± 0.01 | 100 |
| **Starch/Cs (55/45)** | 0.472 ± 0.01 | 75 |
| **Starch/Cs (50/50)** | 0.472 ± 0.01 | 75 |
| **Starch/Cs (65/35)** | 0.417 ± 0.01 | 37,5 |

The browning test is used to ensure plastics bio-composite in this study have a good influence for food preservation. In Table 1 shows that the decreasing weight loss of fruit is smaller than that of the fruit control. Browning index shows that bioplastics for ratio 50/50 + 9% PLM is same for the ratio composition is 55/45 and the lowest browning index is for 65/35 due to the highest amount of starch. As reported in Ref. [3] for low moisture permeability on plastics indicated by the low browning index is very good for food packaging applications. In Ref. [20] also reported that the pore formation is decrease which was consistent with decreasing the humidity levels and also decreasing the browning index [3]. The environmental from the outside are also influenced to the humidity levels when the bioplastic applied for food packaging. Low moisture and low humidity levels correlated with low browning index in this study may due to the effect of amount of PLM in the bio-composite due to the strong bonding formation of C = C, and C-H bond.

## 4. Conclusion

Composite starch/chitosan with the additional Pineapple leaf microfiber has been studied by using FTIR, tensile strength, and XRD for analyzing bonding characteristics, mechanical properties, and structural properties, respectively . The diffraction peak from XRD for composite (starch/chitosan + PLM) shows the new structure was formed as the contribution from the each base materials. FTIR analysis shows the strong bonding formation was formed in bioplastics. Formation are C = C and C-H which was increases with increasing the amount of PLM in composites. Based on the biodegradation test, composites with 9% PLM shows the higher percentage of degradation >80% for 28 days. These results confirm new composite with additional of PLM significantly affect to the decomposition process, contributed to increase the mechanical properties and decrease browning index as an alternative healthy food packaging to maintain freshness.

## Reference

[1] Balakhrisnan P, Sreekala M S, Matjaz K, and Miroslav H (2017) Morphology, transport characteristics and viscoelastic polymer chain confinement in nanocomposites based on thermoplastic potato starch and cellulose nanofibers from pineapple leaf. *Carbohydrate Polymers* **169** pp 176–188.

[2] Dankar J, Amira H, Fawaz E L O, Montserrat P, and Francesc S (2018) Characterization of food additive-potato starch complexes by FTIR and X-ray *Food Chemistry* 260 7–12

[3] Yu Z, Baoqiang L, Jiayu C, and Peifeng Z (2018) Silica in situ enhanced PVA/chitosan biodegradablefilms for food packages. *Carbohydrate Polymers* **184** pp 278–286

[4] Cherian B M, Alcides L L, Sivoney F S, Ligia M M C, Gabriel M O, M Kottaisamy, and Sabu T (2011). Cellulose nanocomposites with nanofibres isolated from pineapple leaf fibers for medical applications. *Carbohydrate Polymers* **86** pp 1790–1798

[5] Mathivanan D, H N Siregar, J P M.R.M R, Bachtiar D, and Cionita T (2016) The study of mechanical properties ofpineapple leaf fibre reinforced tapioca based bioplastic resin composite *ICMER* **74** pp 1–4

[6] Emadian S M, Turgut T O, and Burak D (2016) Biodegradation of bioplastics in natural environments. *Waste Management,* **59** 526–536

[7] Cazon P, Gonzalo V, Jose A, and Manuel V (2016) Polysaccharide-based films and coatings for food packaging: A review. *Food Hydrocolloids* **68** pp 136–148

[8] Sagnelli D, Kim H H, Eric L, Agnes R S, Sophie G, Jacob J K K, Kell M, Denis L, and Andreas B (2016) Plant-crafted starches for bioplastics production *Carbohydrate Polymers* **152** pp 398–408

[9] Ahmed S, and Saiqa I (2016) Chitosan and gelatin based biodegradable packaging films with UV-light protection. *Journal of Photochemistry & Photobiology, B: Biology* **163** 115–124

[10] Costa L M M C, Gabriel M O, Bibin M C, Alcides L L, Sivoney F S, and Mariselma F (2013) Bionanocomposites from electrospun PVA/pineapple nanofibers/ Stryphnodendron adstringens bark extract for medical applications *Industrial Crops and Products* **41** 198–202

[11] Vartiainen J, Mika V N, and Ali H (2014) Biopolymer Films and Coatings in Packaging Applications—A Review of Recent Developments *Materials Sciences and Applications* **5** 708–718

[12] Galaz-E A A, Dalia I S-C, Jaime L C, Ana S S, Tomas J M-S, and Perfecto, P-L (2018) Characterization data of chitosan -based films: antimicrobial activity , thermal analysis, elementary composition , tensile strength and degree crystallinity. *Data In Brief* **21** 473–479

[13] Agustin M B, Bashir A, Shanna M A, and Famille M P (2014) Bioplastic based on starch and cellulose nanocrystals from rice straw. *Reinforced Plastics and Composites* **33** pp 1–9

[14] Brodin M, Maria V, Mihaela T O, Maria C A, and Gary C-C (2017) Lignocellulosics as sustainable resources for production of bioplastics – a review. *Journal of Cleaner Production* **162** pp 646–664

[15] Khan B, Muhammad B K N, Ghufrana S, and Zaib J (2016) Thermoplastic Starch: A Possible Biodegradable Food Packaging Material - A Review. *Journal of Food Process Engineering* **40** pp 1–17

[16] Garcia-Aranda F J, Gonzales R, C F J-G, and E-M (2015) Water Absorption and Thermomechanical Characterization of Extruded Starch/Poly(lactic acid)/Agave Bagasse Fiber Bioplastic Composites. *International Journal of Polymer Science* **2015** pp 609-615

[17] Jabeen N, Ishra, M, and Gulzar N (2015) Bioplastics and food packaging: A review. *Food science & technology* **1** pp 1–6

[18] Mathivanan D B, Siregar J P, Mat Rejab, M R, Bachtiar D, Asmara Y, and Cionita T (2017) The Mechanical Properties of Alkaline Treated Pineapple Leaf Fibre to Reinforce Tapioca Based Bioplastic Resin Composite. *Materials Science Forum* **882** pp 66–70

[19] Wang J, Shenhua S, Shang G, Ravi M, Renchen L, and Qing M (2017) Mg-ion conducting gel polymer electrolyte membranes containing biodegradable chitosan: Preparation, structural, electrical and electrochemical properties. *Polymer Testing* **62** 278–286

[20] Sousa M M A, and Maria P G (2015) Strategies to improve the mechanical strength and water resistance of agar films for food packaging applications. *Carbohydrate Polymers* **132** pp 196–204

**Acknowledgements**

Mr. Heryanto is gratefully acknowledged for his support with the XRD test. This work partially supported by the PDU (Penelitian Dasar Unhas) Grant: 540/UN4.21/PL.00.00/2019 funded by the Hasanuddin University, Makassar Indonesia and PD (Penelitian Dasar) funded by Indonesian Government (Ristek-Dikti).

IOP Conf. Series: Materials Science and Engineering **593** (2019) 012025 doi:10.1088/1757-899X/593/1/012025

# The Formulation of *Apis mellifera* L. Honeybees Feed Making as Pollen Subtitute on Poor Season

**Budiaman[1], M Saleh[2], Yusran[3], S A Paembonan[4], I Gautama [5], D Sanusi[6], S Millang [7]**

[1]Lecturer of Forestry Faculty, Hasanuddin University, Makassar
[2]Assitant of Lecturer in Laboratory of Forest Protection and Benefecial Insects
[3-7]Lecturer of Forestry Faculty, Hasanuddin University, Makassar.

**Abstract**. The Apis mellifera honey bees is one of the most cultured species in Indonesia, both by ordinary people, government (Perum Perhutani and Forest Service) as well as private companies that are managed in a modern way. This research aims to find out the best artificial feed formulation for Apis mellifera hone bees. This research used randomized block design with 4 treatments and 3 replications. The treatment is as follows: Treatment A: Soybean flour + sugar syrup, Treatment B: Dried peanut flour + sugar syrup, C: Peanut flour + sugar syrup, Controls: Natural pollen in flour form. The data obtained were analyzed by Randomized Block Design Equation.

The results showed that the type of pollen substitute TKH and natural pollen is the best feed to increase the growth of Apis mellifera L. bee seedlings in terms of the increase of consumption level and the extent of saplings.

## 1. Introduction

Apis mellifera Bee l. is one of the most cultivated species in Indonesia, both by ordinary people, the government (Perum Perhutani and the Forestry Service) and private companies that are managed in a modern way. A.mellifera bees are the main types of bees that are cultivated in almost all countries, these bees are widely found in Europe [1]. Its ability to produce honey is very high, making many bees introduced to new areas that were previously the deployment of Apis cerana. A. mellifera bee is a honey bee from Italy, but this type of honey bee can adapt well to be developed in Indonesia.

Bee A. mellifera is a honey bee that has high productivity and is easy to adapt to climate in Indonesia [2]. In addition, the specialty of this bee is that it is benign, so it is easily cultivated by the community. Beekeeping can also provide indirect benefits, namely relating to the preservation of forest resources, increasing productivity of crops through mutual symbiosis between plants and honey bees because in the process of finding honey bee food will help the process of pollinating flower plants [3].

Nectar, extranutfialnectar and pollen are foods needed by honey bees for the survival of bees. Nectar is a liquid in plants that is secreted by plants through the nectar gland, which varies in various parts of the plant, nectar is divided into two types, namely floral and extrafloral. Plants that are a source of nectar include Kapuk Randu flowers (Ceiba petandra), Lengkeng (Nephelium longanum), Durian (Durio zibethinus), Sengon (Arachis hypongea) and others. While pollen is a fine material such as powder found on the end of the flower stamen which is a male sex in plants and a source of protein. Honey bees need 20% protein in their food to meet their daily needs [4]. Plants that are the source of pollen include corn (Zea mays), Lamtoro (Leucaena sp), Coconut (Cocos nucifera), Kapuk Randu (Ceiba petandra) and others [1].

The availability of feed in nature is one of the determining factors for the success of honey bee cultivation. Nectar and pollen provide carbohydrates, proteins, fats, vitamins and minerals needed by honey bees for the development of their colonies. Most of the energy needed by honeybees comes

IOP Conf. Series: Materials Science and Engineering **593** (2019) 012025 doi:10.1088/1757-899X/593/1/012025

from nectar, while more protein content is found in pollen needed for larval growth, development of hypopharyngeal glands and body fat [5]. Besides nectar and pollen bees also need.

Bee A. mellifera is a honey bee that has high productivity and is easy to adapt to climate in Indonesia [2]. In addition, the specialty of this bee is that it is benign, so it is easily cultivated by the community. Beekeeping can also provide indirect benefits, namely relating to the preservation of forest resources, increasing productivity of crops through mutual symbiosis between plants and honey bees because in the process of finding honey bee food will help the process of pollinating flower plants [3].

Nectar, extranutfialnectar and pollen are foods needed by honey bees for the survival of bees. Nectar is a liquid in plants that is secreted by plants through the nectar gland, which varies in various parts of the plant, nectar is divided into two types, namely floral and extrafloral. Plants that are a source of nectar include Kapuk Randu flowers (Ceiba petandra), Lengkeng (Nephelium longanum), Durian (Durio zibethinus), Sengon (Arachis hypongea) and others. While pollen is a fine material such as powder found on the end of the flower stamen which is a male sex in plants and a source of protein. Honey bees need 20% protein in their food to meet their daily needs [4]. Plants that are the source of pollen include corn (Zea mays), Lamtoro (Leucaena sp), Coconut (Cocos nucifera), Kapuk Randu (Ceiba petandra) and others [1].

The availability of feed in nature is one of the determining factors for the success of honey bee cultivation. Nectar and pollen provide carbohydrates, proteins, fats, vitamins and minerals needed by honey bees for the development of their colonies. Most of the energy needed by honeybees comes from nectar, while more protein content is found in pollen needed for larval growth, development of hypopharyngeal glands and body fat [5]. Besides nectar and pollen bees also need water to maintain the temperature humidity of the nest, and dissolve honey for consumption [4].

Sidrap Regency is a fairly large corn producing area during the rainy season which is the main source of pollen for honey bee cultivation. Empty land is planted by the community to fulfill poultry feed in Sidrap Regency as the largest egg producer in Eastern Indonesia. But at the end of the rainy season, the availability of pollen is very limited, along with the end of the maize season, so beekeepers need artificial bee feed for the survival of bees farmed.

Limited availability of pollen or non-flowering plants that provide pollen in nature (famine), usually beekeepers overcome by providing pollen substitute feed (Pollen Subtitute) made of materials with high protein content. The problem of pollen availability also arises when farmers farm honeybee colonies in areas that provide a lot of nectar-source plants, but there is no plant source of pollen or pollen, therefore alternative materials are needed to overcome these problems. Some types of nuts have been known to have good nutrients and the percentage of protein is quite high, including soybeans, green beans and peanuts.

## 2. Research Methods

### *2.1 Location and Time*
This research was conducted in March 2017 in Bulo Village, Panca Rijang District, Sidrap Regency, South Sulawesi Province.

### *2.2 Materials and tools*
The equipment used in this study was transparent plastic, digital scale capacity of 1 gram, millimeter paper, scissors, label paper, brushes and stationery. While the materials used in this study were soybeans, green beans, peanuts, natural pollen, sugar and 12 colonies of Apis mellifera bee.

The equipment used in this study was transparent plastic, digital scale capacity of 1 gram, millimeter paper, scissors, label paper, brushes and stationery. While the materials used in this study were soybeans, green beans, peanuts, natural pollen, sugar and 12 colonies of Apis mellifera bees.

### *2.3 Research design*
In this study a randomized block design (RBD) was used with 4 treatments and 3 replications. The treatment s as follows:

a. Treatment A: Soybean flour + sugar syrup
b. Treatment B: Green bean flour + sugar syrup

c. Treatment C: Peanut flour + sugar syrup
d. Control: Natural pollen in the form of flour

*2.4 Variable Observed*
The variables observed in this study were:
a. Replacement feed consumption before and after treatment (gram / week / colony)
b. Area of tillers before and after treatment (cm² / week / colony), after treatment.

*2.5 Data analysis*
The data obtained will be analyzed by group randomized design analysis (RBD) as follows [4]:
Yes i = 1, 2, . . t
J = 1, 2, . . ri
Information:
Yij = Observation value from the first treatment in the jth group
u = Middle population value
i = Additive effect from the i treatment
j = Additive effect from j-group
ij = Effect of experimental error from the first treatment in the j-group
j = test
i = treatment
If the data obtained, the treatment has a significant effect on the observed variables, will be followed by the Tukey follow-up test.

## 3. Result and Discussion

*3.1 Consumption of Substitute Pollen*
Consumption of substitute pollen in colonies that receive various substitute pollen flour has an average range of 29.67-98.67 grams / colony / week. The results of the variance analysis in Appendix 4 show that substitute pollen flour given is significantly different from the consumption of substitute pollen. This might be due to the content of the nutritional value of the feed given is sufficient. According to its function, substitute pollen must fulfill several conditions, namely to attract the attention of honey bees so that honey bees want to eat substitute pollen provided, substitute pollen-making materials are always available in large quantities and cheap, the cost of making replacement pollen is not too large, the nutritional value of the honey fulfills the nutritional needs of honeybees, and does not contain toxic components for bees [12].
The results of the follow-up / Tukey test can be seen in Table 6 as follows:

**Table 1.** Tukey test results Consumption of substitute Pollen.

| Treatments | Average consumption of substitute pollen (grams) | Tukey 0.05 |
|---|---|---|
| TKK | 57,00 | a |
| TKH | 94,67 | b |
| TKT | 29,67 | a |
| POLEN ALAMI | 98,67 | b |

The results of Tukey's further test in Table 1 show that the type of pollen substitute for TKH with an average value of consumption of 97.67 grams / colony / week is the best. The table above also shows that the replacement pollen for TKH with a value of 94.67 grams / colony / week was not significantly different from natural pollen, which was used as a control with a value of 98.67 grams / colony / week. Whereas, for substitute polls for TKK and TKT differ significantly with an average value of 57.00 and 29.67 grams / colony / week against natural pollen with a value of 98.67 grams / colony / week which is used as a control. So it can be concluded that, the type of TKH replacement pollen can replace natural pollen. This is in line with estimates of pollen consumption in worker bees are as much as 3.4 - 4.3 mg per day per head [2]. Feed in the form of pollen is very important for newborn bees, besides pollen also functions to build body fat, the development of the night and ovary glands and extend the life of the bee [2].

IOP Conf. Series: Materials Science and Engineering **593** (2019) 012025 doi:10.1088/1757-899X/593/1/012025

The following diagram shows the average consumption of substitute pollen given to bee A. mellifera for one week as follows:

**Table 2.** Tukey test results Consumption of substitute pollen.

| Treatments | Average consumption of substitute pollen (grams) | Tukey 0.05 |
|---|---|---|
| TKK | 57,00 | A |
| TKH | 94,67 | B |
| TKT | 29,67 | A |
| POLEN ALAMI | 98,67 | B |

The results of Tukey's further test in Table 2 show that the type of pollen substitute for TKH with an average value of consumption of 97.67 grams / colony / week is the best. The table above also shows that the replacement pollen for TKH with a value of 94.67 grams / colony / week was not significantly different from natural pollen, which was used as a control with a value of 98.67 grams / colony / week. Whereas, for substitute polls for TKK and TKT differ significantly with an average value of 57.00 and 29.67 grams / colony / week against natural pollen with a value of 98.67 grams / colony / week which is used as a control. So it can be concluded that, the type of TKH replacement pollen can replace natural pollen [2]

The following diagram shows the average consumption of substitute pollen given to bee *A. Mellifera* for one week as follows

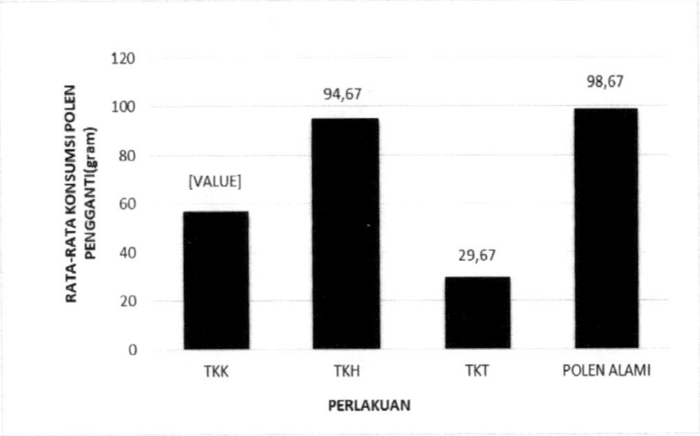


**Figure 1.** Average consumption diagram replacement pollen

Based on Figure 1, in the variable consumption of substitute pollen, the lowest consumption of colonies is a colony that receives TKK and TKT (57 and 29.67 grams / week / colony) while the highest consumption is in colonies that receive natural pollen feed and TKH ( 98.67 and 94.67 grams). The number of substitute pollen consumed by bees tends to be different as seen in Figure 1.

Different consumption of substitute pollen given is thought to be influenced by the texture and aroma of each substitute pollen given. TKT has a slightly sticky and oily texture while TKK has a bad odor and is also a little oily, the texture of TKH is more dense when compared to TKT and TKK. The aroma of artificial pollen TKH has a distinctive aroma of green beans.

The sense of smell in bees was very sharp to find out the aroma of the food he liked with 30,000 odor receptors on the antennae [3]. The most determining factor of pollen (feed) is smell [7].

The low consumption of TKK is also supported by the results of research which provides artificial pollen feed treatment using soybean based ingredients in the form of pasta [6].

*3.2 Area of Tiller Cells*

The extent of tillering cells in colonies that receive various types of replacement pollen, has an average range of 21.56-48.45 cm2 / colony / week. The results of the measurement of the width of bee saplings can be seen in appendix 2. From the data analysis variance in Appendix 4, the replacement pollen given is not significantly different from the total number of tillers cells formed. It is assumed that the nutrient content consumed is not enough. Pollen serves for the development of feeding glands for nurse bees that emit compounds called royal jelly with high protein content and are a food for newly hatched larvae and food for queen bees [8 in 9] .

The following is a diagram of the results of the average cell size of tillers as follows:

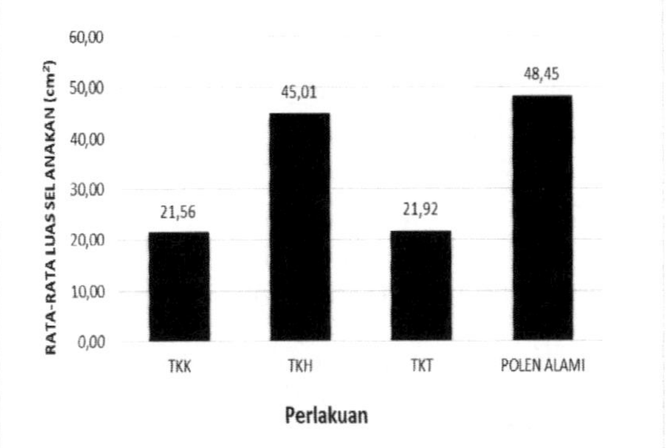


**Figure 2.** Diagram of the average cell size of tillers

Based on the image 2. the lowest area of tillering cells is the colony that gets TKK and TKT (21.56 and 21.92 cm2 / week / colony) while the highest cell size of the tillers is in the colonies that get NATURAL POLEN and TKH (48,45 and 45.01 cm2 / week / colony).

The difference in the extent of tillering cells in the colonies which received various types of pollen substitute flour was suspected, because of the amount of pollen substitute flour consumed. The quantity and quality of pollen in the nest will affect the number of tillers, especially the number of eggs [8]

The large number of bee populations and the extent of tillers in honeybee colonies are also influenced by several factors such as temperature, climate, geographical location, and age of the queen [10]. The factors that cannot be controlled in this study are the fertility of queen bees in laying eggs.

*3.3 Area of Cells Containing Pollen*

The area of pollen-containing cells in the colonies which received various types of pollen pollen substitutes flour, ranged from an average of 31.63 to 81.90 cm2 / colony / week. The results of measurement of cell area containing pollen during the study can be seen in appendix 3. From the variance analysis data based on appendix 4. Flour substitute pollen given is significantly different from the cell area containing pollen. Based on statistical tests, treatments that were significantly different could be tested further by the Tukey test. This can be seen in table 7 below:

**Table 3.** Tukey test results for cell width containing pollen

| Treatments | Average cell area containing pollen (cm2) | **Tukey 0.05** |
|---|---|---|
| TKK | 40,84 | a |
| TKH | 65,46 | b |
| TKT | 31,63 | a |
| POLEN ALAMI | 81,90 | b |

Tukey's advanced test results (Table 3) show the types of pollen substitutes for NATURAL TKH and POLEN provide the best effect on the number of cell areas containing pollen with values (65.46 and 81.90 cm2 / colony / week) compared to substitute polls for TKK and TKT with values (40.84 and 31.63 cm2 / colony / week).

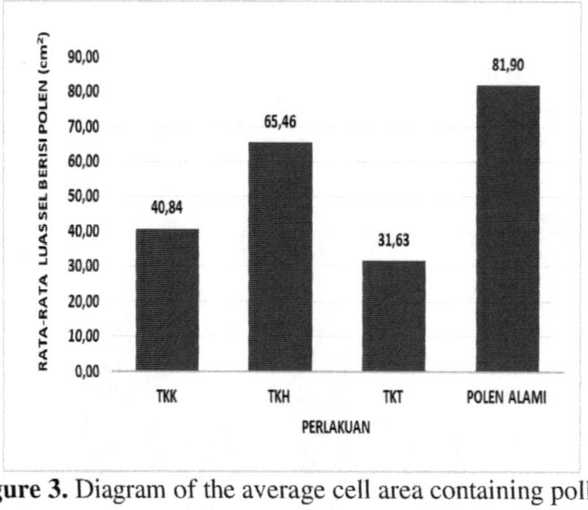


**Figure 3.** Diagram of the average cell area containing pollen

Based on (Figure 3.) showing the lowest pollen cell area is a colony that receives replacement pollen type TKK and TKT with the cell area formed 40.84 and 31.63 cm2 / week / colony, while the highest is in the colony got a replacement pollen type of TKH and NATURAL POLEN with the cell area formed 65.46 and 81.90 cm2 / week / colony. The intensity of worker bees in collecting pollen only reaches the level of fulfilment [8]. In addition, the number of pollen collected will correlate with the number of daughter cells formed or the rate of laying queen bees. This is in line with the number of tillers and the cell area containing pollen formed in this study.

## 4. Conclusion

Based on the results of the research conducted, it can be concluded that the type of pollen substitute for TKH and natural pollen is the best feed for increasing the growth of Apis mellifera L. bee in terms of increased consumption levels and broad tillers.

## 5. Suggestion

*Apis mellifera*, It should be extended past. Respond to flour of various types, possibly only after a few weeks. And avoid using basic language that uses oil.

## References

[1] Pusat Pelebahan APIARI Pramuka 2003 *Lebah Madu: Cara Berternak Lebah dan Pemanfaatan* (Jakarta: Penebar Swadaya)
[2] Adalina Y 2008 *Analisis Finansial usaha Lebah Madu Apis Mellifera L. Jurnal Penelitian hutan dan Konservasi Alam* **5** (3) pp 217–237
[3] Kuntadi 2003 *Perlebahan di Indonesia* (Jakarta: Badan Penelitian dan Pengembangan Kehutanan)
[4] Somerville D 2000 *Honey Bee Nutrition and Supplementary Feeding* (New South Wales: NSW Agriculture)
[5] Winston, M. L. 1987. *The Biologi of Honey Bee*. (England: Harvard University Press)
[6] Agustina D K2008 *Perkembangan Koloni Lebah Madu Apis mellifera L. yang Mendapatkan Pollen Pengganti dari Tiga Jenis Kacang dengan dan Tanpa Vitamin B Komplek* (Bogor: Bogor Agricultural University)
[7] Sihombing D T H 1997 *Ilmu Ternak Lebah Madu*. (Yogyakarta: Gadjah Mada University Press)
[8] Gojmerac W L 1980 *Bees, Beekeeping, Honey dan Pollination* (USA: AVI Publising Co, Inc. Westport, Connectirut)
[9] Sadapotto A 1992 *Pengaruh Variasi Makanan Stimulan terhadap Perkembangan Kapasitas Bertelur Ratu Lebah (Apis mellifera Linn.) di Borongloe, Kecamatan Bontomarannu, Kabupaten Gowa* (Ujung Pandang: Universitas Hasanuddin)
[10] Herbert E W and H Shimunaki 1983 Effect of Mid-season Change in Diet on Diet Consumption dan Bood Rearing By Caged Honey Bees *Apidologie* **14** (2) pp119–125
[11] Minarti S 2010 Ketersediaan Tepung Sari Dalam Menopang Perkembangan Anakan Lebah Madu Apis Mellifera Di Areal Randu (*Ceiba petandra*) Dan Karet (*Hevea brasilliensis*). *Jurnal Ternak Tropika* **11** (2) pp 54–60
[12] Alexander and Klots B E 1961 1001 *Answers to Qustions about Inssects Grossetend Dunlap* (New York)
[13] Gasperz V 1991 *Teknik Analisis Dalam Penelitian Percobaan* (Bandung: Taristo)